Cosmos Creation: The Unified SuperStandard Model, Volume 2, SECOND EDITION

Quantum Entanglement Resolved by a New Wave-Particle Duality,
Precepts for Creating a Cosmos,
Shape of the Universe,
Refined SuperStandard Axioms and Language,
Particle Functional Space, Fourier Wave Space,
Psychology of the SuperStandard Language

Stephen Blaha Ph. D.
Blaha Research

Ex Scienta In Sapientia

Blaha

Pingree-Hill Publishing
MMXVIII

Cover: The cover displays the head of the Titan Prometheus, firegiver to Humanity. Pictures of Newton and Einstein appear on the back cover.

Rev. 00/00/01   July 23, 2018

To My Grandchildren:
Nicholas, Milan, Maxim, and Alexandre

Some Other Books by Stephen Blaha

All the Megaverse! Starships Exploring the Endless Universes of the Cosmos using the Baryonic Force (Blaha Research, Auburn, NH, 2014)

SuperCivilizations: Civilizations as Superorganisms (McMann-Fisher Publishing, Auburn, NH, 2010)

PHYSICS IS LOGIC PAINTED ON THE VOID: Origin of Bare Masses and The Standard Model in Logic, U(4) Origin of the Generations, Normal and Dark Baryonic Forces, Dark Matter, Dark Energy, The Big Bang, Complex General Relativity, A Megaverse of Universe Particles (Blaha Research, Auburn, NH, 2015).

New Types of Dark Matter, Big Bang Equipartition, and A New U(4) Symmetry in the Theory of Everything: Equipartition Principle for Fermions, Matter is 83.33% Dark, Penetrating the Veil of the Big Bang, Explicit QFT Quark Confinement and Charmonium, Physics is Logic V (Blaha Research, Auburn, NH, 2015).

New Boson Quantum Field Theory, Dark Matter Dynamics, Dark Matter Fermion Layer Mixing, Genesis of Higgs Particles, New Layer Higgs Masses, Higgs Coupling Constants, Non-Abelian Higgs Gauge Fields, Physics is Logic VII (Blaha Research, Auburn, NH, 2015)

CQMechanics: A Unification of Quantum & Classical Mechanics, Quantum/Semi-Classical Entanglement, Quantum/Classical Path Integrals, Quantum/Classical Chaos (Blaha Research, Auburn, NH, 2016).

All the Universe! Faster Than Light Tachyon Quark Starships & Particle Accelerators with the LHC as a Prototype Starship Drive Scientific Edition (Pingree-Hill Publishing, Auburn, NH, 2011).

From Asynchronous Logic to The Standard Model to Superflight to the Stars; Volume 2: Superluminal CP and CPT, U(4) Complex General Relativity and The Standard Model, Complex Vierbein General Relativity, Kinetic Theory, Thermodynamics (Blaha Research, Auburn, NH, 2012)

New Boson Quantum Field Theory, Dark Matter Dynamics, Dark Matter Fermion Layer Mixing, Genesis of Higgs Particles, New Layer Higgs Masses, Higgs Coupling Constants, Non-Abelian Higgs Gauge Fields, Physics is Logic VII (Blaha Research, Auburn, NH, 2015)

The Origin of Fermions and Bosons,and Their Unification (Pingree-Hill Publishing, Auburn, NH, 2017).

Megaverse: The Universe of Universes (Pingree Hill Publishing, Auburn, NH, 2017).

SuperSymmetry and the Unified SuperStandard Model (Pingree Hill Publishing, Auburn, NH, 2017).

The Unified SuperStandard Model and the Megaverse SECOND EDITION A Deeper Theory based on a New Particle Functional Space that Explicates Quantum Entanglement Spookiness (Pingree Hill Publishing, Auburn, NH, 2018).

Available on Amazon.com, bn.com Amazon.co.uk and other international web sites as well as at better bookstores (through Ingram Distributors).

CONTENTS

INTRODUCTION..1

1. SOME HIGHLIGHTS OF THE UNIFIED SUPERSTANDARD MODEL 3

   1.1 DERIVATION OF THE UNIFIED SUPERSTANDARD MODEL 3
   1.2 THE MEGAVERSE OF UNIVERSES 6

2. PHYSICAL PRINCIPLES OF CREATION 7

   2.1 CREATION BY A COSMIC PSEUDO-EUCLID ENTITY 7
   2.2 CREATION DYNAMICS RATIONALE..7
   2.3 FUNDAMENTAL PREREQUISITES FOR A FUNDAMENTAL THEORY OF PHYSICS.....................8

3. ENHANCED FORM OF UNIFIED SUPERSTANDARD MODEL AXIOMS 11

   3.1 UNDERLYING BASIS OF SUPERSTANDARD AXIOMS...11
   3.2 PRIMITIVE TERMS AND AXIOMS 11
   3.3 MATHEMATICS AND CONCEPTUAL PREREQUISITES 11
   3.4 PRIMITIVE TERMS FOR THE UNIFIED SUPERSTANDARD MODEL 12
   3.5 REVISED AXIOM SET FOR THE UNIFIED SUPERSTANDARD MODEL..................................13
   3.6 DERIVATION OF THE STRUCTURE OF THE UNIFIED SUPERSTANDARD MODEL 15

4. FACTORIZATION OF QUANTUM FIELDS—WAVE-PARTICLES REALIZED FORMALLY
....17

   4.1 MOTIVATION FOR QUANTUM FIELD FACTORIZATION: INSTANTANEOUS EFFECTS IN QUANTUM
   PHENOMENA..17
   4.2 GENERAL FORM OF FACTORIZATION 17
   4.3 RATIONALE FOR FACTORIZATION 18
   4.4 WAVE-PARTICLE DUALITY MATHEMATICALLY REALIZED 18
   4.5 VISUALIZATION OF A PARTICLE..19
   4.6 FACTORIZATION OF FERMION QUANTUM FIELDS 20
      4.6.1 Mass of a Qube..20
      4.6.2 Qube Spin...21
      4.6.3 Qubes as Fermion Field Functionals...21
      4.6.4 Dirac-like Equations of Matter from 4-Valued Logic 23
      4.6.5 Functional Expression for Each of the four Species of Fermions 24
   4.7 FACTORIZATION OF BOSON QUANTUM FIELDS..24
      4.7.1 Quba Cores of Fundamental Bosons...25
   4.8 TYPES OF FUNCTIONALS FOR FERMIONS AND BOSONS 25

5. FUNCTIONAL SPACE OF PARTICLES AND INTERNAL SYMMETRY GROUPS.............27

   5.1 PARTICLE FUNCTIONAL SUBSPACE..27
   5.2 FUNCTIONAL COORDINATES AND INTERNAL SYMMETRY GROUPS 27
   5.3 FUNCTIONAL TRANSFORMATIONS THAT LEAD TO FUNCTIONAL INTERNAL SYMMETRY GROUPS 28

5.4 SUBSPACES OF PARTICLE FUNCTIONAL SPACE (A UNIVERSE) 28
5.5 DISTANCE IN PARTICLE FUNCTIONAL SPACE..29
5.6 PRODUCTS OF FUNCTIONALS – FUNCTIONAL ALGEBRA 29
5.7 TRANSFORMATIONS BETWEEN FUNCTIONALS 29
5.8 FUNCTIONAL INTERNAL SYMMETRY GROUP TRANSFORMATIONS..29
5.9 ANALOGY BETWEEN COORDINATE AND PARTICLE FUNCTIONAL SPACES 30

6. THE WAVE SPACE OF FREE FOURIER EXPANSIONS...31

6.1 SUBSPACES OF WAVE SPACE (UNIVERSE) 31
6.2 INTERNAL QUANTUM NUMBERS OF WAVE FOURIER EXPANSIONS 31
6.3 DISTANCE IN WAVE SPACE 32
6.4 SPACE-TIME TRANSFORMATIONS AMONG WAVE FOURIER EXPANSIONS.................................32
6.5 INTERNAL SYMMETRY TRANSFORMATIONS OF WAVE SPACE FOURIER EXPANSIONS 32

7. PARTICLE FUNCTIONAL-LAGRANGIANS 33

7.1 SKELETON FUNCTIONAL LAGRANGIANS 33
7.2 FUNCTIONAL-LAGRANGIAN AND FEYNMAN DIAGRAMS 34
7.3 PRODUCTION RULES FOR FUNCTIONAL-LAGRANGIANS...34

**8. COMPUTATIONAL LANGUAGE INTERPRETATION OF PARTICLE FUNCTIONAL
TRANSFORMATIONS..35**

8.1 LANGUAGES AND GRAMMARS...35
8.2 EXAMPLE OF PRODUCTION RULES...36
8.3 EXAMPLE OF THE PRODUCTION RULES FOR A LAGRANGIAN INTERACTION TERM....................37
8.4 PARTICLE FUNCTIONAL TRANSFORMATIONS IDENTIFIED AS PRODUCTION RULES 38

9. QUANTUM FIELD THEORY WITH WAVE-PARTICLE FUNCTIONAL FORMALISM......39

10. THE SHAPE OF UNIVERSES AND SURFACE TENSION 41

10.1 SHAPE OF A UNIVERSE..41
10.2 SPHEROIDAL UNIVERSES 42

11. THE PSYCHOLOGY OF PHYSICAL REALITY...43

11.1 PROGRESS IN UNDERSTANDING OF PHYSICAL REALITY 43
11.2 LANGUAGE AND PSYCHOLOGY...43
11.3 ENTITY DECISION AXIOMS 44
C.1.3 Decision Axioms for a Fundamental Theory 44
    C.1.3.1 Ockham's Razor..44
    C.1.3.2 Leibniz's Minimax Principle for Physics Theories 44
    C.1.3.3 Conputationally Minimal 45
    C.1.3.4 Experimentally Verifiable Directly or Indirectly...45
    C.1.3.5 Level of Rigor 45
    C.1.3.6 Replication Principle - Nature Tends to Repeat Successful Strategies....................45
    C.1.3.7 Principle of Maximal Serendipity 46
    C.1.3.8 Principle of Maximal Descent to a Complete Theoryy 46

11.4 Principle of Vitamorphism 46
11.5 Wave-Particles vs. Thought 46

**12. PROPOSALS FOR THE FUTURE OF ELEMENTARY PARTICLE THEORY AND
EXPERIMENT** **47**
12.1 Theory 47
12.2 Experiment Proposals 47
    12.2.1 Speed of Light *47*
    12.2.2 Extremely Massive Particles *47*
    12.2.3 Dark Particles *48*
    12.2.4 Left-Right Imaginary Momenta of Quarks *48*

REFERENCES **49**

INDEX **55**

ABOUT THE AUTHOR **57**

FIGURES and TABLES

Figure 4.1 Symbolic view of a free particle having particle functional and fourier expansion parts. The 'smearing' of the particle by the Two-Tier Y field described in the Book is symbolically displayed.. 19

Table 4.1 Core functionals within the various types of fundamental elementary particles. 25

Figure 5.1. Schematic for the procedure for generating a functional Internal Symmmetry group U_f from sets of U transformations of a functional 4-vector f_μ... 28

Figure 5.2. The set of four layers of internal symmetry groups in Particle Functional Space corresponding to the four layers of spin ½ fermions and the four layers of vector bosons. In addition there are the Species group and the Interaction Rotations group Θ. 30

INTRODUCTION

This is volume 2 of the Second Edition of *The Unified SuperStandard Model and the Megaverse*, which will be called the ***Book*** in the text of this volume. This volume provides additional material on topics covered in the Book as well providing a revised, more compact, set of axioms that are equivalent to those stated in the Book. The purpose of defining a revised, slightly expanded, set of axioms is to sharpen our understanding of the fundamental constructs of the theory and to emulate the simplicity of Euclid's axioms of Geometry as much as possible.

The volume begins with a description of the guiding physical principles of the creation of the Cosmos of universe(s). Then an enhanced set of axioms are presented that lead to the derivation of the Unified SuperStandard Model given in the Book.

Next it defines wave-particle duality with quantum fields expressed as inner products of particle functionals and fourier wave expansions. This form of wave-particle duality resolves the problem of the instantaneity (spookiness) of entangled quantum states.

The volume then describes the universe of particle functionals and the universe of fourier wave expansions first presented in the Book and its predecessor edition. Particle functional lagrangians for functional transitions are described, which specify production rules, and are then put within the framework of type 0 Chomsky languages.

The impact of particle functionals on perturbation theory is explained. The language view of the Unified SuperStandard Model leads to a consideration of the physical psychology it embodies.

We also show that the surface tension of a universe implies it has the shape of a spherical 3-surface. We conclude by suggesting a needed series of elementary particle experiments.

1. Some Highlights of the Unified Superstandard Model

1.1 Derivation of the Unified SuperStandard Model

The Book and earlier books have developed a comprehensive derivation of the known features of elementary particles as embodied in The Standard Model, and extended in The SuperStandard Model, to include a much richer spectrum of fundamental fermions and bosons. The major omission in the derivation is the specification of particle masses and interaction coupling constants. An interesting point that emerged in the consideration of coupling constants, including the gravitational constant, was, when expressed using a form of coherent 'vacuum' state expressions, all coupling constants had values of the order of unity. This fact suggests that *bare* values of the coherent vacuum state coupling constants have the value one. Interactions modify the values to those observed in experiments.

In this section we will <u>briefly</u> summarize highlights of the derivation of the Extended SuperStandard Model presented in the Book:

1. The number of spatial dimensions was determined to be the number of generators in the primary set of interactions of the space. In the case of an *empty* universe the primary set of iteractions is the U(2) qubit transformations group. The number of U(2) generators is four and thus the dimension of space is 4 complex dimensions. Also and more importantly, considerations of Asynchronous Logic, and the requirement that physical processes must be able to proceed in parallel, require the number of spatial dimensions to be four. The Book justifies four complex space-time dimensions with a Lorentz metric yielding Complex Lorentz group symmetry.

2. Boosts of the Complex Lorentz group transform a Dirac-like equation with a Landauer mass into four different forms (called species). Each form maps to a type of fermion: neutral leptons (neutrinos), charged leptons, up-type quarks, and down-type quarks. Neutral leptons and down-type quarks are tachyons. Some evidence exists for tachyonic neutrinos. Complex Lorentz boosts lead to the Complex Lorentz group factorization: $SU(2) \otimes U(1) \otimes SU(3) \otimes SU(2) \otimes U(1)$. We map $SU(2) \otimes U(1) \otimes SU(3)$ to fermion particle functional space to obtain the internal symmetry group for ElectroWeak and Strong Interactions: $SU(2) \otimes U(1) \otimes SU(3)$. The remaining factors $SU(2) \otimes U(1)$ we map to the internal symmetry group for Dark Matter, which we take to be the Dark ElectroWeak Interaction (unconnected to normal matter interactions) .

3. The existence of four conserved (and partially conserved) quantum numbers such as baryon number and lepton number indicates that there is a U(4) group whose 4 representation causes each species to have four generations—three of the generations are known. We suggest that a fourth generation of much higher mass fermions exists.

4. In each generation there are four partially conserved quantum numbers. Thus we find that there is another U(4) group (called a Layer group) for each generation yielding the combined Layer groups $[U(4)]^4$. The 4 representation of each U(4) results in a fermion spectrum of four layers of four generations or 192 fermions in all. We see only one layer at present. The additional three layers of fermions remain to be found at much higher masses. The symmetry group of the Unified SuperStandard Model is

$$[SU(2){\otimes}U(1){\otimes}SU(3){\otimes}SU(2){\otimes}U(1){\otimes}U(4){\otimes}U(4)]^4{\otimes}U(4)$$

where the last factor is for the broken Species group, which follows from Complex General Relativity..

5. Assuming all particles are massless at the Big Bang, and all particle types have an equal proportion of the total mass-energy then, we find that the 192 fermions and 192 vector bosons yield a Dark Matter percentage of 83.33% (experimentally the estimates are 84.5% and 81.5%). The proportion of Dark Mass-Energy is found to be 91% of the universe's mass-energy. Experimentally the proportion has been estimated to be 95%. These results agree well with experiment. See chapter 14 of the Book for details.

6. The instantaneity of quantum effects between space-like separated parts of a quantum state ('spookiness') is taken to be a feature of fundamental importance. The only sensible way to implement this feature in quantum theory is to assume that the wave function of every particle is the inner product of a particle functional and a fourier coordinate expansion. Particle functionals exist in a space with no distance measure. The space of coordinate fourier expansions also has no distance measure. Other functionals in a state (and their implicit coordinate fourier expansions) change *instantaneously* when one of the functionals comprising a state changes since coordinate space distance is irrelevant.

7. Fermion particle functionals are called *Qubes*. They exist 'within' every fermion. They have a mass that we take to be the Landauer mass—the minimal energy of a qubit. Boson particle functionals are called *Qubas*. They are assumed to be massless in the absence of all interactions to preserve free vector boson aand spin 2 boson gauge symmetry. Free Higgs particles are assumed to be massless for consistency.

8. To have a completely finite theory with no infinities (including no fermion triangle infinities) we introduced Two-Tier Coordinates that replaced normal pointlike coordinates with a kind of 'fuzzy' coordinates.

$$X^\mu = x^\mu + iY^\mu(x)/M_c^2.$$

9. Since the Unified SuperStandard Model lagrangian would require higher order derivatives to account for quark confinement (linear potential terms) and for MOND-like deviations from conventional gravity, and since such terms would be outside a canonical lagrangian formulation, we introduced two fields for each particle (fermions and bosons) in a formulation we call Pseudoquantum Theory. Pseudoquantum theory enables a canonical lagrangian formulation. It has other advantages such as a clean separation of vacuum expectation values from quantum fields for Higgs particles. It also supports second quantization in arbitrary coordinate systems while maintaining the same particle interpretation of states in all coordinate systems.

10. The Book also describes Higgs symmetry breaking and the use of the Faddeev-Popov Mechanism in detail for the theory.

11. Since a Complex Special Relativity requires a Complex General Relativity we considered Complex General Relativity and showed that it could be 'factored' into General Relativity and a new U(4) group that we called the Species group. Since Complex General Relativity must support interactions with all types of matter we specified a Species group interaction with all matter. Further, we assumed that the Species vector bosons acquired masses through the Higgs Mechanism The Higgs Mechanism caused Species group contributions to each fermion mass. Such a mass term would require each fermion particle mass to be both inertial *and* gravitational *solving the mystery of the equality of inertial and gravitational mass.*

12. We showed that the implicitly higher derivative Riemann-Christoffel curvature tensor for all interactions leads to new interactions beyond The Standard Model. In addition to yielding quark confinement and MOND-like modifications of gravity, it may help understand the missing nucleon spin issue, discrepancies in proton radius measurements, vector meson dominance (VDM), and so on.

13. We defined an Interaction Rotations group that caused rotations among all the vector boson interactions of The Extended SuperStandard Model. We found that rotations that respected Superselection rules such as the Charge Superselection rule could have physical significance. One example is ElectroWeak Theory which is an application of Interaction Rotation transformations.

14. Since the number of fundamental fermions (192) and fundamental vector interaction bosons (192) is equal we considered Supersymmetric-like features of the Extended SuperStandard Model.

1.2 The Megaverse of Universes

The Book describes a Megaverse of universes. Highlights of this discussion include:

1. A discussion of evidence for universes beyond our universe.

2. Arguments for the dimension of the Megaverse to be 192,[1] the number of vector bosons of the Unified SuperStandard Model interactions:

$$[SU(2) \otimes U(1) \otimes SU(3) \otimes SU(2) \otimes U(1) \otimes U(4) \otimes U(4)]^4$$

(absent the Species group vector bosons).

3. The universes within the Megaverse have surfaces that exhibit a form of *surface tension*. The surface tension force on a universe is inward directed. It is the combined effect of the gravity generated by all mass-energy within the universe—just as the surface tension of a water droplet is generated by the attractive inter-molecular forces of the interior water molecules.

4. Universes treated as a type of particle called universe particles are described and their dynamics described.

5. General features of the Megaverse including Megaverse Quantum Field Theory are described in some detail.

---

[1] The number of vector boson fields is 192 for this group product. The number of generators of the U(192) Interaction Rotations group is 192^2.

2. Physical Principles of Creation

2.1 Creation by a Cosmic Pseudo-Euclid Entity

The derivation of the Extended SuperStandard Model seems to be best presented within the framework of a putative being who may or may not be a 'real' entity. Of this 'being', which might have been called the Cosmic Pseudo-Euclid Entity, but we will call the Entity, we can say: 1) It must be 'outside of' (independent of) time, 2) It must be 'immaterial' (not composed of anything), 3) It must be 'unchanging' in itself and 4) It originates the derived theory presented here and in the Book.

In the Book we have called this Entity the 'Unmoved Mover' that implements the theory as Reality and causes all events to happen in a manner consistent with the Unified SuperStandard Model.[2]

In this chapter we start with a discussion of the basic prerequisites of Creation[3] (a general theory of everything) and then proceed in chapter 3 to specify slightly revised axioms of creation[4] that still yield the derivation presented in the Book.

2.2 Creation Dynamics Rationale

The derivation of the theory for our universe and for other possible universes[5] within a Megaverse, has certain fundamental prerequisites if one wishes to have dynamical physical processes, as we know them, to occur within interesting universes. We take these prerequisites for granted normally and proceed to concoct theories of physics as if they may be pulled out of a hat like the proverbial rabbit. However any attempt to create a universal physical theory must meet these prerequsites to build a theory from fundamental primitive terms and axioms in a manner similar to Euclid's construction of geometry.

---

[2] The role and features of the Entity would suggest that it is God to many. However, since we address only Physics issues, we shall leave its nature as an open issue not resolved by Physics.

[3] Non-physical Aside: One might ask What may motivate an omniscient being with limitless power to create from Nothingness? The only answer apparent to the author is *Loneliness*—A feeling not uncommon in humanity. Note that the point of Creation does not have a time since time is an artifact of Creation.

[4] The axioms presented here are more or less equivalent to those of the Book. In the author's view they are simpler and cleaner.

[5] We will call other universes *exoverses*.

2.3 Fundamental Prerequisites for a Fundamental Theory of Physics

We can list fundamental prerequisites based on a general knowledge of the necessary nature of a fundamental theory of Physics. This approach presumes a general knowledge of the theory that we wish to construct illustrating the maxim, "Our ends determine our beginnings."

1) A time variable must exist that may have various forms,

2) We wish to have a dynamical fundamental theory that evolves in time. Thus there must be a mechanism(s) that allow dynamical processes to exist that may, or may not, run in parallel.

3) Multiple parallel processes can execute.

4) There must be a space with a coordinate system(s), and distance measure, within which processes can execute.

5) There must be particles upon which dynamical processes execute.

6) There must be a space of functionals that support the creation of particle states and help determine their properties. The particle functional space frees particles from a complete dependence on the coordinate space.

7) There must be a space of 'waves' of free field fourier expansions for all the fundamental particles absent interactions..

8) There must be an order in the 'created' dynamical theory that is embodied in a form of a computational language[6] with a Chomsky-like *Grammar* using a finite set of terminal and nonterminal *symbols* that constitutes an alphabet (vocabulary).[7] The ordering in the form of a language with grammar *Production Rules* ensures the consistency of the generated theory.[8]

---

[6] The possibility that the universe is one enormous Word was explored in *Cosmos and Consciousness* (Blaha (2003)) in physical, philosophical, and religious contexts. A few years ago around 2012 the author found a book with a similar title by R. M. Bucke published in 1901 entitled *Cosmic Consciousness* on the evolution of Man to a new level of consciousness. The content of this book is inrelated to Blaha (1998) – first edition - and (2003) as well as Blaha's other books.

[7] Particle Computer Languages are described in Blaha (2005b) and (2005c) as well later in this volume and in other books by the author.

[8] See chapter 8 for definitions and details.

9) Creation should opt for Vitamorphic[9] universes that support life in some form. Recent studies have shown that evolution favors the development of increasingly intelligent life. Thus the ultimate appearance of intelligent life at places within universes appears to be natural making the *Anthropic Principle* an evolutionary consequence[10] of the *Vitamorphic Principle*.

These prerequisites would seem to be necessary and sufficient for the specification of primitive terms and axioms for a fundamental theory.

---

[9] The *Vitamorphic Principle* states that universes should support some form of life realizing that there are many varieties of life and borderline forms of life. A 'tight' definition of life has not been satisfactorily constructed. There are many borderline entities that may or may not be called life. We take 'Vitamorphic' to mean 'life enabling' in English. Vitamorphism is not a concept without meaning—a universe (Megaverse) consisting of only inert matter without energy present would be non-Vitamorphic. The Anthropic Principle, briefly put, states that intelligent human-like life should exist.

[10] One can well wonder whether the emergence and dominance of Mankind has eliminated the possibility of the emergence of other intelligent species on earth from the many semi-intelligent species that exist now and in the past.

3. Enhanced Form of Unified SuperStandard Model Axioms

3.1 Underlying Basis of SuperStandard Axioms

In this chapter we present a revised set[11] of 'primitive' terms and axioms for our theory. A comparison of this new set of axioms with those provided in the Book will show that they are equivalent but add a few new axioms. They are also more simply stated, have fewer overlaps between axions, and cleanly lead to the Book's theory of elementary particles.

The goal of the Book and this volume is to derive the Unified SuperStandard Model in the manner of Euclid with a clear connection between the steps of the derivation just as Euclid developed geometry from a progression of theorems.

3.2 Primitive Terms and Axioms

Primitive terms can be as simple as those of Euclid or they can be more complex. The level of simplicity depends on the nature of the theory and the Physical Laws that emerge from it. In the case at hand, a fundamental unified theory, the constructs that emerge in the construction of the theory are mathematically complex. Consequently, the choice of primitive terms and axioms may be expected to be mathematically complex as well, unless one wishes to expand the primitive terms into a more detailed, term by term description in simpler, more basic primitives. We will not pursue that alternative here since the terms that we use are 'self-explanatory' to the Elementary Particle Physics theorist knowledgable about quantum field theory and particle symmetries.

3.3 Mathematics and Conceptual Prerequisites

Due to the complexity of the Theory we have chosen to specify mathematics prerequisites and use them in the derivation rather than devoting parts of the derivation to mathematical preliminaries. Therefore we use complex variable theory, Riemannian coordinates, group theory, classical and Quantum Logic, functionals, Chomsky-like computational languages, and so on without bringing in unnecessary supporting details from them.

We also assume certain physical concepts such as distance, quantum features, second quantization, covariance under a group transformation, and spatial curvature.

---

[11] The Book presents axioms in chapter 0.

The list of axioms uses some of these prerequisite concepts treating them as primitive terms for the derivation.

3.4 Primitive Terms for the Unified SuperStandard Model

The somewhat revised set of primitive terms of the theory are:

> Qubits
> Qubes
> Qubas
> Core
> Grammar
> Terminal and Nonterminal Symbols
> Production Rules
> Speed of Light
> Spatial Dimensions
> Space and Time Coordinates
> Covariance under group transformations
> Asynchronous processes
> Parallel Processes
> Reference Frame
> Complex Lorentz Group
> General Coordinate Transformations
> Gravity
> Universe
> Particle Masses
> Fermions
> Bosons
> Particle States
> Particle Rest State
> Particle Momenta
> Spin
> Canonical Quantization
> Quantum Process
> Quantum Entanglement
> Second Quantization
> Quantum Field Theory
> Quantum States
> Asymptotic Particle States
> Internal Symmetries
> Coupling Constants

> Discrete Symmetries
> Yang-Mills Local Gauge Theory
> Functionals
> Functional space

In choosing these primitives, we understand that they generally embody a significant theoretic description or body of knowledge. We do not include names used in the mapping to reality (such as quark) in the list of primitives since the mapping to reality is a separate issue in our view.

3.5 Revised Axiom Set for the Unified SuperStandard Model

The somewhat revised set of axioms that we list below is supplemented by the Decision Axioms of Appendix C.1.3 of the Book. The 'new' physical axioms are

PARTICLE AXIOMS
1. All matter and energy is composed of particles.
2. Each fundamental particle has a physico-logic structure within it that we designate its core.
3. Particles form an alphabet with a finite number of characters and combine in ways specified by the quantum probabilistic production rules of a quantum computational grammar.[12]
4. A core is a particle functional that combines with a free field fourier coordinate expansion in an inner product to produce a free second quantized particle field.
5. There is a 4-dimensional space of particle functionals, called *particle functional space*, with the distance measure eq. 3.1 specifying the transformation group of particle functionals.
6. Particle functional space consists of a single point.
7. The core of a fermion functional is called a *qube*. Fundamental bosons have a core consisting of a boson functional called a *quba*.
8. Qubes have a a a bare mass. Qubas have zero mass.

SPACE AXIOMS
9. The dimensions of a coordinate space-time are determined by the number of fundamental[13] interactions, and the requirement that all parallel processes, with parts perhaps separated by distances, can occur synchronously.
10. Spatial coordinates are inherently complex-valued.
11. Space has one complex-valued component that plays the role of time. Physical phenomena dynamically evolve based on the time variable.

---

[12] See Blaha (2005b).
[13] Interactions that would exist in the absence of fermion particles.

12. The infinitesimal distance ds between two space-time points is given by

$$ds^2 = dt^2 - d\mathbf{x}^2 \tag{3.1}$$

where $d\mathbf{x}$ is a vector of the spatial coordinates. Transformations between coordinate systems preserve the value of ds and define a transformation group. (The Complex Lorentz Group)

13. Physically acceptable reference frames have real-valued coordinates. These coordinates can be obtained by group transformations from complex-valued coordinate systems. Physical space-time measurements are made in a real-valued coordinate system.

14. The speed of light is the same in all reference frames.

15. Free fundamental leptons must have a real-valued energy.

16. Gravity may cause space-time to be curved. (Complex General Coordinate transformations[14])

DYNAMICS AXIOMS

17. The complete theory has a lagrangian formulation. If the lagrangian is truncated to quadratic form (interactions set to zero) then symmetries appear that are the source of particle symmeytry groups that persist with broken symmetry after interactions are reintroduced. The lagrangian specifies a set of production rules of a type 0 Chomsky language generalized to include production rules for the generation of all strings of symbols (particles) from any strings of symbols (including the *head symbol*.)[15]

18. The lagrangian of the theory must be invariant under coordinate system transformations.

19. Dynamical particle equations must be covariant under group transformations.

20. All interactions have a local Yang-Mills gauge theory formulation.

21. The vector bosons, and the interactions among them, are determined by terms in complete lagrangian, some of whose parts are obtained from the Riemann-Christoffel Curvature Tensor.

QUANTIZATION AXIOMS

22. All fields must be canonically quantized.

23. Fermion and Boson vacua can be defined that are valid in all coordinate systems.

---

[14] If the metric tensor of space-time is analogous to one of the metric tensors of the superfluid phases of ^{3}He, then space-time might have several metric tensors in 'various regions.' If the space-time metric tensor is analogous to the ^{3}He-B superfluid phase metric tensor, which has an effective gravity with a complex metric tensor, the space-time metric tensor would be the familiar one of General Relativity. However if the space-time metric tensor is analogous to the metric tensor of superfluid ^{3}He-A, which exists at higher pressure and temperature, then the space-time metric tensor might be similar to the Penrose twistor theory metric tensor. In this case the corresponding General Relativity may have a twistor-like metric tensor: perhaps in the early universe, and/or inside black holes, and/or in small universes with higher pressure and temperature than our universe. We will assume the conventional metric for Complex Special and General Relativity.

[15] Chapter 8 discusses computational languages for particles in detail.

24. The number of particles in an asymptotic state of any given type is invariant in all reference frames.
25. Quantum processes starting in an initial quantum state, with parts separated by a distance after a time, can have the parts synchronously change each other instantaneously. (Quantum Entanglement)

3.6 Derivation of the Structure of the Unified SuperStandard Model

The derivation provided in the Book follows from the above axioms as well as from the axioms stated in the Book. The axioms in the Book are a subset of the axioms in section 3.5.

4. Factorization of Quantum Fields—Wave-Particles Realized Formally

Many years ago Dirac factored the Klein-Gordan equation and obtained the Dirac equation for spin ½ fermions. In the Book and in this chapter we show that there is good reason to factor quantum mechanical wave functions and second quantized fields into an inner product of a particle functional and a corresponding fourier coordinate expansion. With this factorization, and the assumption that the space of all particle functionals, as well as the space of all fourier coordinate expansions, are both located at a point with the consequence that there is no distance measure in either space, we find a change in one of a pair of space-like separated parts of an initial state causes an instantaneous transformation of the other part (Einstein's spookiness). Because of the significance of this result we have implemented the factorization of wave functions and quantum fields as axioms.

This chapter describes factorization for fermions and bosons. Much of it appears in the Book.

4.1 Motivation for Quantum Field Factorization: Instantaneous Effects in Quantum Phenomena

Instantaneous quantum phenomena are apparent in many cases. For example:

1. Two particles placed in a definite spin state can separate to a space-like distance. If the z component of spin is flipped in one of the particles, the other particle instantaneously flips its spin in such a way as to conserve spin. This type of phenomena has been described as 'spooky' since it violates the law that no effect can travel at a rate faster than the speed of light.

2. Transitions between atomic levels take place instantaneously—in a zero time interval.

4.2 General Form of Factorization

Normally fermion and boson quantum fields are described by a wave function of the form

$$\chi(\mathbf{x}, t) \qquad (4.1)$$

We can formally factorize quantum fields as an inner product of a functional f_k and a space-time fourier expansion denoted (k, **x**, t) (neglecting internal quantum numbers temporarily) wher k is the momentum.

$$\chi(\mathbf{x}, t) = (f_k, (k, \mathbf{x}, t)) \tag{4.2}$$

For a free two particle wave function (non-interacting) the wave function may be written as a product of inner products:

$$\chi(\mathbf{x}, t) = (f_{1k}, (k, \mathbf{x}, t)_1) (f_{2q}, (q, \mathbf{x}, t)_2) \tag{4.3}$$

where k and q are momenta.

4.3 Rationale for Factorization

The rationale for factorization lies in the nature of the functionals and coordinate fourier expansions that we use. For, we choose to create a space of particle functionals for fermions and bosons that consists of a single point with no distance measure (or alternately put, zero distance between all functionals.) We also choose to create a 'point' space of all coordinate fourier expansions for bosons and fermions, whose elements have all coordinate values, x. We will discuss these space in more detail in chapters 5 and 6.

For the moment we wish to note that the space of functionals consists of functionals for all fundamental particles in the universe (Megaverse). Chapter 7 describes transitions (interactions) in which functionals are transformed into other functionals. So the space of functionals has a dynamic aspect. Another important aspect of functional space is its universality—*all functionals of the Megaverse are present creating a type of link between all parts of the Cosmos.*

The space of coordinate fourier expansions consists of all possible expansions for particles in the coordinates of each respective universe and of the Megaverse. This space also has no distance measure.

The factorization that we propose, as exemplified by eqs. 4.2 and 4.3, enables instantaneous communication of a transition between two space-like separated parts of a state. A change in one part immediately causes a corresponding change in the other part because the changes take place in the functionals which are located at the same point in functional space.

In a certain sense we have divorced quantum phenomena from coordinate space by quantum field factorization.

4.4 Wave-Particle Duality Mathematically Realized

At the beginning of Quantum Theory in 1924, De Broglie postulated wave-particle duality and made progress in understanding quantum phenomena. Wave-particle duality was subsequently 'abandoned' in favor of the Quantum Mechanics of Heisenberg and Schrödinger.

The quantum field factorization that we propose provides a mathematical formulation of particles that separates the particle part (the functional) from the wave part (the coordinate fourier expansion.) Thus we have realized wave-particle duality for the purpose of understanding instantaneous effects of quantum entanglement.

However, in our formulation, quantum fields $\chi(x, t)$ appear in the SuperStandard lagrangian. *Then* perturbation theory calculations of phenomena are made using free field fourier expansions, denoted (x, t), for all fermions and bosons. Thus we achieve a quantum theory (non-deterministic) and yet have wave-particle duality embedded within it (unlike De Broglie-Bohm theory.)

4.5 Visualization of a Particle

We can visualize an elementary particle located at point x as composed of two components: a functional f and a fourier expansion denoted by x. They follow from the above discussion and the derivation in The Book:

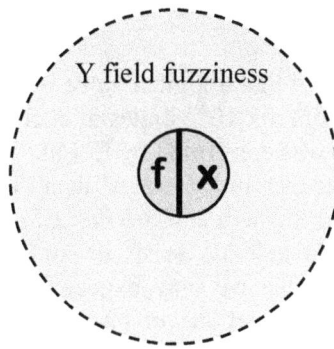

Figure 4.1 Symbolic view of a free particle having particle functional and fourier expansion parts. The 'smearing' of the particle by the Two-Tier Y field described in the Book is symbolically displayed.

The central disk represents the core of the particle which is located at x but 'smeared' by the cloud of Y particles of Two-Tier Quantization. Thus point particles do not exist in the full SuperStandard Model theory and thus infinities are not encountered in the calculation of any diagram in perturbation theory.

The central disk (which is not necessarily truly centered) represents a Qube functional for fermions and and a Quba functional for bosons. (See below.) As described in The Book Qubes embody spin ½ and Qubas embody integer spin. Qubes have a bare mass m_0 and Qubas are massless. The bare mass of Qubes reflects the fact that fermions have spin and thus an intrinsic energy while Qubas can carry energy and thus spin while in motion but elementary bosons act only as 'delivery agents' for spin (and thus quantum values).

4.6 Factorization of Fermion Quantum Fields

If we consider all possible 'things' that might constitute a fundamental building block for a fundamental fermion theory they are all, at best, *ad hoc* and raise questions of their necessity and whether they are composed of a yet more fundamental substructure.

There is only one choice of building block that avoids these issues – a logic unit or qubit. A qubit is a fundamental entity that is a complex form of computer bit. A bit (and thus a qubit) is known to have an energy or equivalently a mass, and has no constituents of a more primitive form.[16] We call a unit of logic that forms the 'core' of a particle a *qube*.[17] It exists as the core of a fermion particle. But, in itself, it has no *independent* material existence or space-time coordinates. A qube is a functional that acquires features such as coordinates, to become an elementary particle. We define a qube as a fermion field theory functional. (See chapters 3 and 8 of the Book.) We now introduce physical features that will cloak qubes with properties and interactions making them into fundamental fermion quantum fields.

4.6.1 Mass of a Qube

Recent experiments have shown that a logical value of a qubit has an energy associated with it. One bit of information has about 3×10^{-21} joules of energy[18] or a rest mass, m_0, or about 0.02 eV using $m_0 = E/c^2$. This result was confirmed by E. Lutz et al.[19] who showed that there is a minimum amount of heat produced per bit of erased data. This minimal heat is called the *Landauer*[20] *limit*. The equivalent mass we will call the *Landauer mass* and denote it as m_0. We will assume that a fundamental Landauer mass exists in our discussions although the precise value of the mass will not be used since we may expect all physical particle masses to be renormalized to different values when interactions are taken into account.

We will assume all fermions contain a qube within them. A qube is assumed to have mass m_0. The masses of fermions are modified to their known values by interactions.

It is intriguing that the mass of the electron neutrino has been measured in a variety of experiments and found to be within an order of magnitude or so larger than our estimate of the Landauer mass (as we would expect since particles acquire a 'cloud of virtual particles' due to interactions.) This 'cloud' can be expected to increase its mass above the Landauer mass. Since neutrinos only have the weak interaction it is not surprising that the increase due to interactions

---

[16] A qube is a physical manifestation of a logical value. The relation of a qube to a logical value is analogous to the relation of a penciled point placed on paper to the concept of a point as a primitive in geometry.

[17] In the First Edition we called qubes iotas. However, since the name iota was previously used as a particle name many years ago it seemed reasonable to use a different name. We chose the name 'qube' for self-evident reasons. *'Qube' is pronounced 'cube.'*

[18] E. Muneyuki et al, *Nature Physics*, DOI: 10.1038/NPHYS1821.

[19] E. Lutz et al, Nature **483** (7388): 187–190,10.1038/nature10872, (2012).

[20] R. Landauer, "Irreversibility and heat generation in the computing process", IBM Journal of Research and Development **5** (3): 183–191, (1961).

should not be large. The Mainz Neutrino Mass Experiment, for example, estimates the electron neutrino mass to be less than 2 eV.

A number of astronomical studies have also generated estimates of neutrino masses. In July 2010 the 3-D MegaZ DR7 galaxy survey found a limit for the combined mass of the three neutrino varieties to be less than 0.28 eV.[21] A smaller upper bound for the sum of neutrino masses, 0.23 eV, was found in March 2013 by the Planck collaboration,[22] In February 2014 a new estimate of the sum was found to be 0.320 ± 0.081 eV due to discrepancies between the Planck's measurements of the Cosmic Microwave Background, and other predictions, combined with the assumption that neutrinos are the cause of weaker gravitational lensing than implied by massless neutrinos.[23]

Thus the experimentally measured values of neutrino masses are consistent with the qube Landauer mass estimate of 0.02 eV given above. We thus assume that *a fermion particle consists of an qube with a certain mass,*[24] *that is renormalized, together with other features. These features emerge in the derivation of the complete theory in the Book.*[25]

We view Reality as ultimately a representation (or painting) of logic values evolving through interactions in time and space.[26]

4.6.2 Qube Spin

The spin of a qube is assumed to be spin ½. Qubes are solely a building block of fermions.

4.6.3 Qubes as Fermion Field Functionals

At this point qubes have an insubstantial appearance with only the attributes of mass and spin. Later we will suggest, in detail, that they can be mathematically represented as fermion field functionals[27] and used to develop the structure of the fermion spectrum and The Unified SuperStandard Model.

---

[21] S. Thomas et al, "Upper Bound of 0.28 eV on Neutrino Masses from the Largest Photometric Redshift Survey", Physical Review Letters **105**: 031301 (2010).

[22] Planck Collaboration, arXiv:1303.5076 (2013).

[23] R. A. Battye et al, "Evidence for Massive Neutrinos from Cosmic Microwave Background and Lensing Observations", Phys. Rev. Lett. **112**, 051303 (2014).

[24] Leibniz first proposed the idea of logic 'particles' which he called monads. Our definition of a logic 'particle' does not include (or exclude) the presence of a spiritual part which was part of the definition of Leibniz's monads.

[25] A recent experiment claims to separate the spin part (which we identify as a logical value later) of a molecule from the rest of the molecule.

[26] Those who might suggest matter is substantial, and logic values are not, should remember that matter would be completely insubstantial if there were no forces in nature. Neutrinos which are close to insubstantial would be completely insubstantial if there were no weak interactions.

[27] Functionals are a mathematical primitive of our theory. They have been used extensively by Feynman and others in quantum theories.

We will see that the SuperStandard Model interactions and features such as Quantum Entanglement *require* the use of a functional formalism for particle fields.

The deeper formulation of particle qubes supports the theory presented in the First Edition and Second Edition.

Now we define a canonical functional approach to creating a simple Dirac fermion quantum field from a qube and a fourier quantum expression for the space-time part of a free fermion quantum field. We symbolize a qube for a fermion with no internal symmetry and spin ½ as f. We begin by defining a coordinate space Dirac fourier quantum expansion as

$$(p, s, x, t) = N(p)[b(p, s)u(p, s)e^{-ip \cdot x} + d^\dagger(p, s)v(p, s)e^{+ip \cdot x}]$$

where $N(p)$ is a normalization factor, u and v are functions of spin and momentum, and b and $d^\dagger$ are creation/annihilation operators.

A Dirac quantum wave function can be defined as an inner product of a qube functional and a coordinate space fourier quantum expansion. For example

$$\psi(x) = (f_p, (p, s, x, t)) = \sum_{\pm s} \int d^3p N(p)[b(p, s)u(p, s)e^{-ip \cdot x} + d^\dagger(p, s)v(p, s)e^{+ip \cdot x}] \qquad (4.4)$$

where we use a functional inner product formalism in the manner of Riesz (1955)[28] and others. Note, p is a symbol in the inner product—not the integration variable on the right.

A functional inner product yields a numeric value. In the present case, it yields a numeric (possibly quantum) function. In general an inner product of a functional f with a variable function g is expressed as

$$G(x) = (f, g(x)) \qquad (4.5)$$

For each value of x, $G(x)$ has one numeric value modulo quantum smearing.

We can use the above inner product expression for a fermion field for each of the four general types of fermions presented in the Book:[29] a Dirac type of fermion, a tachyonic type of fermion, a Dirac type of fermion with a complex 3-space momentum fourier expansion; and a tachyonic type of fermion with a complex 3-space momentum fourier expansion.Thus there are four differing functionals initially – one for each of the four types of fourier expansions. Internal symmetries, introduced in the Book, will lead to a multi-dimensional space of functionals. We discussed these points in detail in chapter 3 of the Book.

---

[28] For example see pp. 61-2 of Riesz (1955) where linear functionals and their inner products are defined.

[29] These types of fermion fields can be further subdivided into left-handed and right-handed fields.

4.6.4 Dirac-like Equations of Matter from 4-Valued Logic

In our derivation every truly fundamental particle of matter, whether quark or lepton, has spin ½. We have seen in chapter 10 of Blaha (2011c) that the basic algebra of Operator Logic eigenvalue operators, and that of its raising and lowering operators, are the same as the algebra of creation and annihilarion operators for free spin ½ particles. Our goal is to build our theory on the scaffolding of Operator Logic. We view a fermion particle as a qube core which is dressed in spatial coordinates (and internal symmetries):

$$\text{Qube core} + \text{coordinates} \rightarrow \text{fermion particle} \qquad (4.6)$$

The creation and annihilation operators $b(p,s)$ and $d^\dagger(p,s)$ (and their hermitean conjugates $b^\dagger(p,s)$ and $d(p,s)$) are mathematically similar to the raising and lowering operators of Operator (Matrix) Logic. They satisfy the anticommutation relations

$$\{b(q,s), b^\dagger(p,s')\} = \delta_{ss'}\delta^3(\mathbf{q} - \mathbf{p}) \qquad (4.7)$$
$$\{d(q,s), d^\dagger(p,s')\} = \delta_{ss'}\delta^3(\mathbf{q} - \mathbf{p})$$

Thus we see spin ½ particle wave functions originating from the Dirac-like spinors, and raising and lowering operators of the spinor formulation of Operator Logic.

When particles interact the quantum field theory interaction terms use fermion creation operators, $b(q,s)$ *and* $d^\dagger(q,s)$, *and annihilation operators,* $b^\dagger(p,s')$ *and* $d(q,s)$, *to implement the transformations between the Qubes of the interacting particles.*[30] *Thus the mathematics of the embedded Qubes' logic values is automatically implemented within quantum field theoretic calculations.*

An interesting point that emerges from this discussion is the nature of spin ½ particle states such as

$$|p, s> = b^\dagger(p, s)|0> \qquad (4.8)$$

This state is interpreted as a one particle state. It also has an analogous interpretation in Operator Logic as creating a one term universe of discourse – a construct which is in part linguistic and in part logic. Thus particles are embodiments of Logic values and particle interactions change the logic values of the initial particles to those of the emergent particles. All in all, our universe can be viewed as an extraordinarily intricate logic machine. Serendipitously we are now seeing the use of particles to create quantum computers, which, in a sense, is bringing us full circle. Particles are Logic; Logic machines emerge from particle interactions.

---

[30] See chapter 3 of the Book.

4.6.5 Functional Expression for Each of the four Species of Fermions

We have derived the four species of fermions in the Book. We have used a 'conventional' notation for quantum fields. In this section we will define these quantum fields as inner products of functionals and fourier coordinate expansions.

Dirac Quantum Field:
$$\psi(x) = (_1f, \text{ Dirac\_fourier\_expansion})$$

Tachyon Quantum Field:
$$\psi_T(x) = (_2f, \text{ Tachyon\_fourier\_expansion})$$

Complexon Quantum Field:
$$\psi_C(x) = (_3f, \text{ Complexon\_fourier\_expansion})$$

Complexon Tachyon Quantum Field:
$$\psi_{CT}(x) = (_4f, \text{ Tachyon\_Complexon\_fourier\_expansion})$$

The digit prefixes of $_kf$ for k = 1, 2, 3, 4 distinguish the functionals for each species.

In addition we can decompose the above quantum fields into left-handed and right-handed fields. The left-handed functional representations are:

Left Dirac Quantum Field:
$$\psi_L(x) = (_{1L}f, \text{ left-handed\_Dirac\_fourier\_expansion})$$

Left Tachyon Quantum Field:
$$\psi_{TL}(x) = (_{2L}f, \text{ left-handed\_Tachyon\_fourier\_expansion})$$

Left Complexon Quantum Field:
$$\psi_{CL}(x) = (_{3L}f, \text{ left-handed\_Complexon\_fourier\_expansion})$$

4.7 Factorization of Boson Quantum Fields

The First Edition of the Book opened the possibility that fermions (and bosons) might have a core that embodies logic in the form of spin as well as bare masses in the case of fermions. In the Second Edition of the Book we expanded our discussion to better describe the functionals' space, within which the core functionals of each type of fundamental particle reside.

We defined functionals of various elementary boson spins: 0, ½, 1, and 2. Bosons have cores as well that are bosonic functionals with integer spin. We called a bosonic core a *quba*[31] in analogy with the fermion functionals name of qubes. Bosonic functionals are massless. Bosons acquire masses through interactions. The rationales for logic cores for particles was discussed in detail in chapters 3 and 8 of the Book.

4.7.1 Quba Cores of Fundamental Bosons

We define a corresponding boson functional quba for each type of elementary boson. We will designate a boson functional as b_s where s specifies the spin which may be 0, 1, or 2. Every boson contains a boson functional core within it. A quba has the spin of the elementary boson within which it resides. It has zero mass since bosons are typically massless prior to symmetry breaking effects. The space of Quba functionals is described in chapter 5.

4.8 Types of Functionals for Fermions and Bosons

The functional content embodied in each type of elementary particle is summarized in Table 4.1.

PARTICLE TYPE	CORE	MASS	SPIN
Fermion	qube	m_0	½
Scalar Boson	quba	0	0
Vector Boson	quba	0	1
Graviton	quba	0	2

Table 4.1 Core functionals within the various types of fundamental elementary particles.

---

[31] We use 'quba' simply because of its similarity to 'qube'. The leading 'b' signifies its bosonic use. We pronounce 'quba' as 'bub' with a silent 'e.' The word 'quba', itself, is the name of a Bantu language spoken by the Bubi people of Bioko Island in Equatorial Guinea.

5. Functional Space of Particles and Internal Symmetry Groups

5.1 Particle Functional Subspace

The functionals of elementary particles form a space[32] that includes all the free field fermion and boson functionals of our universe and any other universe that might exist (the Megaverse). All fundamental fermions and bosons have a corresponding particle functional. Fermion particle functionals f… are labeled with momentum k, internal symmetry quantum numbers denoted λ, and possibly other subscripts ζ such as handedness, and PseudoQuantum field type (1 or 2). Boson particle functionals b… are labeled with momentum k, spin s, internal symmetry quantum numbers denoted λ, and possibly other subscripts ζ such as handedness, and PseudoQuantum field type (1 or 2):

$$f_{k\lambda\zeta}$$
$$b_{ks\lambda\zeta} \qquad\qquad (5.1)$$

5.2 Functional Coordinates and Internal Symmetry Groups

Functional Space also includes generic quadruplets of functionals that we can treat as functional coordinates of a 4-space of functionals as we did in the Book. These functional coordinates can be transformed by Complex Lorentz transformations to be coordinates of a different coordinate system.

Further, as section 3.3 of the Book shows, one can create sets of coordinate systems that can be mapped to representations of all the subgroups of the Complex Lorentz group: SU(3), SU(2), U(1), SU(2), and U(1). These groups, defined by maps of sets of representations, can combine to form a functional Reality group with commuting factors for Internal Symmetries: R = SU(3)⊗SU(2)⊗U(1)⊗SU(2)⊗U(1). In addition functional coordinates can be transformed by U(4) transformations to be coordinates of sets of coordinate systems that can be mapped to U(4) representations of a functional Generation group, the Layer groups, and the Species group as well as the group of the Interaction Rotations group U(192). Thereby we define the set of commuting functional Internal Symmetry groups of Fig. 5.1 which comprise the groups of the Unified SuperStandard Model: $[SU(3)\otimes SU(2)\otimes U(1)\otimes SU(2)\otimes U(1)\otimes U(4)\otimes U(4)]^4\otimes U(4)$

Operators of the functional groups that we define, transform fundamental particles among each other.Thus our Functional Space contains both functional Internal Symmetry

---

[32] Much of this chapter appears in the Book.

groups and the fundamental particle functionals upon which they operate.The functional Internal Symmetry groups can be dressed in space-time coordinates to yield the Internal Symmetry groups of the Unified SuperStandard Model.The fundamental particle functionalscan be 'dressed' in space-time coordinates using functional inner products as described earlier and in the Book.

5.3 Functional Transformations that Lead to Functional Internal Symmetry Groups

The procedure, dscribed in section 3.3 of the Book, for generating functional Internal Symmmetry groups from sets of transformations of a functional 4-vector is somewhat involved. In this section we will diagram the procesdure for creating a functional group U_f of the form of one of the factors of

$$[SU(3)\otimes SU(2)\otimes U(1)\otimes SU(2)\otimes U(1)\otimes U(4)\otimes U(4)]^4\otimes U(4) \tag{5.2}$$

from a generic functional 4-vector f_μ.

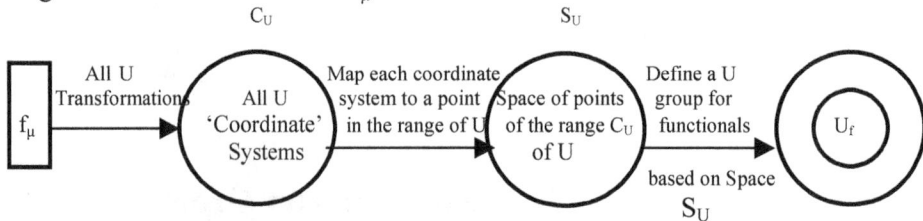

Figure 5.1. Schematic for the procedure for generating a functional Internal Symmmetry group U_f from sets of U transformations of a functional 4-vector f_μ.

Applying this procedure for each factor of eq. 5.2 gives a set of groups that commute with each other. From these functional groups we obtain the symmetry groups of the Unified SuperStandard Model. Also the symmetry groups, that we obtain, have no difficulties with No Go theoorems such as the Coleman-Mandula Theorem.

An alternate approach, based on the above procedure, is to define each group factor of eq. 5.2 in functional space in such a way that the defined functional implementations of the group factors commute, and then to use them to define fermion and boson particle functionals of the form of eq. 5.1 as eigenstates of subsets of factors.

5.4 Subspaces of Particle Functional Space (a Universe)

Particle subspaces of the particle functional space can be categorized by spin: spin ½ functionals, spin 0 functionals, spin 1 functionals, and spin 2 functionals. Each subspace has elements for all physical values of k, s, λ, and ζ.

5.5 Distance in Particle Functional Space

There is no distance in particle functional space—or otherwise stated, all distances are zero and the space consists of a single mathematical point.

5.6 Products of Functionals – Functional Algebra

Due to the Feynman propagators that appear for all particles in Dyson-Wick perturbation theory calculations[33] (see chapter 9 below), which contain quantum fields (the inner products of functionals and fourier expansions) the product of two functionals is not defined. One can assume

$$f_{k\lambda\zeta}f_{k\lambda\zeta} = 1$$
$$b_{ks\lambda\zeta}b_{ks\lambda\zeta} = 1 \qquad , \qquad (5.2)$$

or not. We will treat particle functionals as commuting c-number quantities

$$[\,_if_\xi, \,_{i'}f_{\xi'}] = [\,_if_\xi, b_{s'\xi'}] = [b_{s\xi}, b_{s'\xi'}] = 0 \qquad (8.9)$$

as we did in the Book.

5.7 Transformations between Functionals

The Unified SuperStandard Model lagrangian has numerous interaction terms. Each interaction term defines a transformation between particle functionals in perturbation theory diagrams. Chapter 9 discusses particle transformations in the perturbation theory of quantum field theories. Chapter 7 shows how a lagrangian can be put in a form that gives functional transitions. Chapter 8 shows that these transformations can be put in the form of language grammar rules.

5.8 Functional Internal Symmetry Group Transformations

The λ subscript contains the indices of various fundamental representations of internal symmetry groups: $[SU(3)\otimes SU(2)\otimes U(1)\otimes SU(2)\otimes U(1)\otimes U(4)\otimes U(4)]^4$ where the last two U(4) factors are for the Generation groups and the Layer groups.[34] Fig. 5.2 shows the internal symmetry groups of Particle Functional Space that 'rotate' particle functionals.

Since Particle Functional Space does not have a coordinate space dependence the internal symmetry groups are independent of coordinate space and evade 'No Go' theorems of

---

[33] F. J. Dyson, Phys. Rev. **82**, 428 (1951); G. C. Wick, Phys. Rev., **80**, 268 (1950); See also chapter 6 of Blaha (2007b), and chapter 17 of Bjorken (1965).
[34] The U(4) Species group, resulting from Complex General Relativity (broken) is assumed to have a singlet representation for all particles 1 since it appears to be the only reasonable choice for a symmetry originating in General Relativity. All particles should have a unique interaction arising from Relativity and should not be gravity quadruplets.

the 1960's. (The dummy 'k' index is merely to indicate that an integration over momentum takes place in functional inner products.)

5.9 Analogy Between Coordinate and Particle Functional Spaces

In chapter 8 of the Book we discussed the relation of Particle Functional Space Internal Symmetries and the coordinate space Reality group.   We found an analogous relationship between the Reality group of coordinate space and of corresponding groups in Particle Functional Space. W*e have found a new infinite pointlike space—Particle Functional Coordinate Space.*

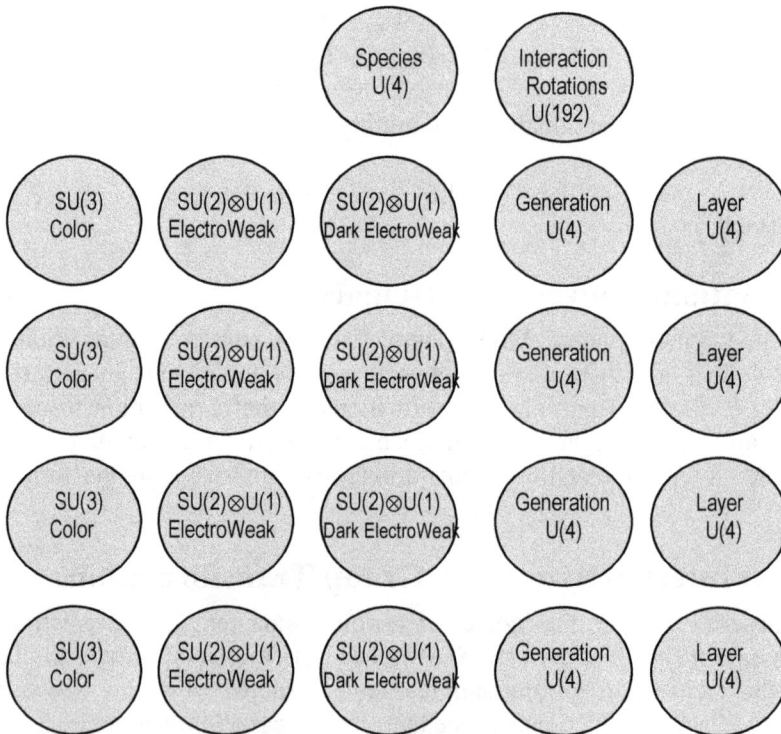

Figure 5.2. The set of four layers of internal symmetry groups in Particle Functional Space corresponding to the four layers of spin ½ fermions and the four layers of vector bosons. In addition there are the Species group and the Interaction Rotations group Θ.

Functional Space contains the functionals of all fermion and boson particles, and the functionals of all fields of the groups of the Unified SuperStandard Model.

6. The Wave Space of Free Fourier Expansions

The set of *free*[35] fourier wave function expansions for all fundamental fermion and boson fields forms a space[36] that we call *Wave Space*. The fourier wave function expansions contain creation and annihilation operators. Therefore Wave Space is a q-number space with commuting wave function expansions and non-commuting fourier wave function expansions as the case may be.

Wave Space supports Lorentz transformations of the space-time coordinates and momenta, and Internal Symmetry transformations as well.

6.1 Subspaces of Wave Space (Universe)

Wave Space can be divided into subspaces according to their spin and internal quantum numbers. The spin subspaces are for spin 0, ½, 1, and 2. Within spin subspaces there are subspaces corresponding to combinations of the internal symmetry group factors of eq. 5.2.

6.2 Internal Quantum Numbers of Wave Fourier Expansions

We can label fourier wave function expansions as

$$(s, \mathbf{x}, t)_{k\lambda\zeta} \qquad (6.1)$$

where the labels are: momentum k, internal symmetry quantum numbers denoted by λ, and possibly other subscripts ζ such as handedness. For example, a fermion fourier expansion has the general form

$$(s, x, t)_{k\lambda\zeta} = N_{\lambda\zeta}(k)[b_{\lambda\zeta}(k, s)u_{\lambda\zeta}(k, s)e^{-ik\cdot x} + d_{\lambda\zeta}^{\dagger}(k, s)v_{\lambda\zeta}(k, s)e^{+ik\cdot x}] \qquad (6.2)$$

where s = ½, and a free field defined by the functional inner product

---

[35] We take a perturbation theory view of the Unified SuperStandard Model, and other quantum field theories. The free fields obtained from functional inner products are used in perturbation theory to calculate quantities and processes of interest. Thus our restriction to free fourier expansions does not impede the determinations of perturbation theory results.
[36] Much of this chapter appears in the Book.

$$\psi_{\lambda\zeta}(x) = (f_{k\lambda\zeta}, (s, x, t)_{k\lambda\zeta}) = \sum_{\pm s} \int d^3k N_{\lambda\zeta}(k)[b_{\lambda\zeta}(k, s)u_{\lambda\zeta}(k, s)e^{-ik\cdot x} + d_{\lambda\zeta}^{\dagger}(k, s)v_{\lambda\zeta}(k, s)e^{+ik\cdot x}]$$

$$(6.3)$$

6.3 Distance in Wave Space

We assume the space-time distance between fourier wave function expansions to be *zero* in keeping with the zero distance between particle functionals in functional space. This assumption is solidly based on the instantaneity of transformations of parts of entangled states. (No spookiness!) Separating the parts of a quantum state S into space-like separated parts S_1 and S_2.we find a change in one part causes an instantaneous change in the other part:

$$<x|S> \rightarrow <x_1|S_1><x_2\|S_2>$$

$$(6.4)$$

irrespective of distance since the implicit functionals and fourier expansions within each has no space-time separation from each other.

6.4 Space-Time Transformations Among Wave Fourier Expansions

Space-time Lorentz and Poincaré transformations of the coordinates and momenta can be applied to Wave Fourier Expansions.

6.5 Internal Symmetry Transformations of Wave Space Fourier Expansions

Internal Symmetry transformations can be applied to Wave Space Fourier Expansions using the indices within λ and ζ.

7. Particle Functional-Lagrangians

7.1 Skeleton Functional Lagrangians

If we could imagine a 'snapshot' of the universe[37] taken at one instant of time we could presumably enumerate all the functionals of the universe's particles. Then succeeding snapshots would show an ebb and flow of functionals as time progresses. This thought brings us to the important issue of the transformations of particle functionals in particle interactions. The simplest statement that one could make about functional transformations is that they are created and annihilated according to the interaction terms of the 'skeletonized' Unified SuperStandard Model [38] We call 'skeletonized' lagrangians *Functional-Lagrangians*.

We skeletonize a lagrangian density by deleting all quadratic terms and then replacing all fermion and boson particle fields in interaction terms by their corresponding functionals.[39] For example the lagrangian

$$\mathcal{L} = \bar{\psi}_C (i\gamma^\mu D_\mu - m)\psi_C(x) + b(\bar{\psi}_C \psi_C(x))^2 \tag{7.1}$$

becomes the skeleton lagrangian

$$\mathcal{L}_S = bf^4 \tag{7.2}$$

where f is the fermion's functional. We delete derivative operators

$$\partial_\mu \psi_{\lambda\zeta} \rightarrow f_{\lambda\zeta} \tag{7.3}$$

where $f_{\lambda\zeta}$ is the functional corresponding to $\psi_{\lambda\zeta}(x)$. We also delete γ-matrices and factors of i.

---

[37] We realize that such a snapshot is not possible since infinite velocity particles that could feed a camera this snapshot do not exist.

[38] Functional lagrangiaans are discussed in chapter 8 of the Book.

[39] In our construction of particle functional space we have not introduced complex conjugation of functionals for lack of a good reason. Complex conjugation takes place only in the fourier expansion part of a quantum field. Another issue is the appearance of lagrangian terms with factors that are drivatives of fields. Since we do not do computations with skeleton lagrangians we can ignore the derivative in each such factor and simply substitute the functional. For example, $\varphi^3(\partial^\mu\varphi)^2$ becomes the quba expression b^5.

Thus our skeletonized lagrangian formalism only describes the transitions between functionals in an interaction. This formalism is made more concrete by considering Feynman diagrams for the interactions.

7.2 Functional-Lagrangian and Feynman Diagrams

Feynman diagrams with their in and out ordering specify the transformations between functionals more completely. A simple example shows the interaction transformations of functionals. Consider the lagrangian term

$$(\bar{\psi}\psi(x))^2(\partial^\mu\varphi)^2$$

One corresponding Feynman diagram for it is

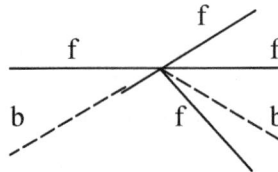

with qube functionals labeled f and quba functionals labeled b.

In the above example we have not introduced internal symmetries. When internal symmetries are introduced then the skeletonized lagrangians and the corresponding Feynman diagram representations are significantly more complicated.

7.3 Production Rules for Functional-Lagrangians

Each term in a Functional-Lagrangian represents a set of language production rules. Sections 8.3 and 8.4 present an example of production rules so generated. The mapping to language production rules illustrates the discussions in chapters 1 and 2 concerning the language interpretation of the Unified SuperStandard Model.

8. Computational Language Interpretation of Particle Functional Transformations

In this chapter[40] we will discuss a language interpretation of particle functional transformations based on a Chomsky-like language and grammar. We will see that particle functional transformations can be viewed as grammar production rules with the net result that the evolution of the universe (Megaverse!) can be viewed as the evolution of an enormous Word consisting of a very large but finite number of (terminal) symbols.

8.1 Languages and Grammars

In chapter 3 of Blaha (2005b) we described a linguistic interpretation of particle interactions. In this interpretation particles play the role of symbols (terminal symbols and nonterminal symbols[41]) in an alphabet (of a finite number of symbols) for a Chomsky-like language. Chomsky defines four types of language ranging from type 0 through type 3. Particle theories, when viewed in terms of their perturbation expansions, can be viewed as a generalization of a type 0 language. A type 0 language (also called an unrestricted rewriting system) allows any grammar production rule of the form

$$X \rightarrow Y$$

where X and Y are sets of particles (strings).

A production rule is a specification of a transformation of a string of symbols (set of particles) to another string of symbols (set of particles). In the case of quantum theories production rules are inherently quantum probabilistic.

A grammar is specified by a quadrupole of items symbolized by the expression

---

[40] Most of this chapter appeared in Blaha (2005b) and other books by the author. For more than a hundred years mathematicians and physicists have been describing Physics and Mathematics as being a language in colloquial, layman's terms. Our books show that elementary particle physics, such as our SuperStandard Model, is precisely a type 0 Chomsky language, in which production rules are generated from lagrangian interaction terms. Specific examples are presented in Blaha (2005b) and (2005c).

[41] From a particle view a terminal symbol is a particle that appears in input or output states (strings) of a perturbation theory diagram. A nonterminal symbol is a particle that appears in an intermediate state of a perturbation theory diagram. In the theories that we have considered particles are both terminal and nonterminal symbols.

<N, T, S, P>

where N is a set of nonterminal symbols, T is a set of terminal symbols, S is a special terminal symbol called the head or start symbol, and P is a finite set of production rules. In quantum theories such as the SuperStandard Model N and T coincide. The start symbol S corresponds to the bare vacuum. Chomsky's definition of a language is the set of all strings of terminal symbols that can be generated from the start symbol using the production rules. *We extend the definition of a particle language to the set of all finite strings of particles (symbols) whether or not they can be generated from the start symbol.*[42]

The set of production rules is finite in Chomsky's definition of language. In the context of quantum field theories we note that a lagrangian of the form of a finite polynomial expression is equivalent to a finite set of production rules.

8.2 Example of Production Rules

In this section we will consider a simple example of production rules with the alphabet:

Start Symbol: S
Nonterminal symbols: A, B
Terminal symbols: x, y

We choose the production rules:

$$S \rightarrow AB \qquad \text{Rule I}$$
$$A \rightarrow y \qquad \text{Rule II}$$
$$A \rightarrow Ay \qquad \text{Rule III}$$
$$B \rightarrow x \qquad \text{Rule IV}$$
$$B \rightarrow Bx \qquad \text{Rule V}$$

An example: Generating a string ('particles') yyxxx from the head symbol S using the above production rules:

$$S \rightarrow AB$$
$$AB \rightarrow AyB$$
$$AyB \rightarrow yyB$$

---

[42] If one considers the fact that all particles originate either directly or indirectly from the Big Bang (the Start symbol), then the Chomsky definition of type zero languages applies where all strings originate in the Start symbol of the Big Bang.

$$yyB \rightarrow yyBx$$
$$yyBx \rightarrow yyBxx$$
$$yyBxx \rightarrow yyxxx$$

8.3 Example of the Production Rules for a Lagrangian Interaction Term

Earlier we noted that a particle lagrangian with a finite number of terms polynomial in particle fields terms would always have a corresponding finite set of production rules. In this section we consider the example of a lagrangian electromagnetic interaction term for electrons and positrons:

$$e\gamma \cdot Ae$$

This lagrangian term yields the production rules:

$$e \rightarrow eA$$
$$e \rightarrow Ae$$
$$eA \rightarrow e$$
$$Ae \rightarrow e$$
$$p \rightarrow pA$$
$$p \rightarrow Ap$$
$$pA \rightarrow p$$
$$Ap \rightarrow p$$
$$ep \rightarrow A$$
$$pe \rightarrow A$$
$$A \rightarrow ep$$
$$A \rightarrow pe$$

where e represents an electron, p represents a positron, and A represents the electromagnetic field. Blaha (2005b) presents sequences of transitions using the above production rules and their corresponding Feynman-like diagrams.

Blaha (2005b) also presents other examples such as the ElectroWeak Interaction:

$$\nu_e W^- e$$

where v_e is an electron type neutrino, and W^- is a negative Weak W vector boson.

8.4 Particle Functional Transformations Identified as Production Rules

In chapter 7 we showed how to define a lagrangian for functionals that provided transformation rules for functionals. In this section we consider the example of a functional lagrangian electromagnetic interaction term for electron and positron functionaals:

$$f_e \gamma \cdot b_A f_e$$

This functional lagrangian term yields the functional production rules:

$$f_e \rightarrow f_e b_A$$
$$f_e \rightarrow b_A f_e$$
$$f_e b_A \rightarrow f_e$$
$$b_A f_e \rightarrow f_e$$
$$f_p \rightarrow f_p b_A$$
$$f_p p \rightarrow b_A f_p$$
$$f_p b_A \rightarrow f_p$$
$$b_A f_p \rightarrow f_p$$
$$f_e f_p \rightarrow b_A$$
$$f_p f_e \rightarrow b_A$$
$$b_A \rightarrow f_e f_p$$
$$b_A \rightarrow f_p f_e$$

where f_e represents an electron functional, f_p represents a positron functional, and b_A represents the electromagnetic field functional.

9. Quantum Field Theory With Wave-Particle Functional Formalism

Quantum field theory calculations are almost always done in perturbation theory. Perturbation theory expansions[43] use vacuum expectation values of time ordered products of pairs of quantum fields. Since quantum fields, in our functional wave-particle formulation, are the result of inner products of functionals (particle cores) and waves (fourier wave expansions), the form of quantum field vacuum expectation values is the same as usually found.

Therefore our new deeper level of our understanding of particle structure does not change perturbation theory. However it does account for the instantaneity of effects in separated parts of a quantum entangled process.

---

[43] See chapter 6 of Blaha (2007b), and chapter 17 of Bjorken (1965) for formulations of perturbation theory.. Bjorken (1965) and Blaha (qqqaaa).

10. The Shape of Universes and Surface Tension

10.1 Shape of a Universe

The shape of a universe[44] either in the case of a Cosmos consisting of one universe or a Megaverse of universes has been a subject of much interest. A universe can be viewed as a 3-surface within a larger space of dimension D > 3.[45]

One interesting possibility, a Poincaré dodecahedral shape, was proposed by J. P. Luminet et al.[46] Experimental evidence from WMAP appears to have ruled out this case.[47]

In this chapter we suggest the surface tension of a universe, a topic introduced in the Book, may make a universe a spherical 3-surface.[48] The Book showed that there was an inward directed surface tension force on the surface of a universe due to gravitation between the matter and energy within the universe.

In the absence of other nearby universes, we expect the 3-surface of the universe to be spherical if a uniform distribution of mass-energy in the large is assumed—a common assumption. If there are other nearby universes then the 3-surface of the universe might remain spheroidal but no longer 'perfectly' spherical.

The example of spherical droplets of water in a uniform environment provides a perfect analogy to the effect of surface tension on the shape of a universe. In the case of a water droplet the inward directed surface tension is caused by intermolecular forces between water molecules.

Given a spherical universe one can calculate its surface area and volume. Blaha (2017c), to which the reader is referred, contains estimates of the radius of our expanding universe in various models.

---

[44] See Blaha (2004).

[45] In the case: of D = 4 the map of the 3-surface has been done by R. Lehoucq, J. Weeks, J.-P. Uzan, E. Gausmann, and J. P. Luminet, arXiv:gr-qc/0205009 (2002).

[46] J. P. Luminet, J. R. Weeks, A. Riazuelo, and J.-P. Uzan, Natue **435**, 593 (2003)..

[47] N. J. Cornish, D. N. Spergel, G. D. Starkman, and E. Komatsu, Phys. Rev. Lett. **92**, 201302 (2004).

[48] We note that there appears to be spherical symmetry at the Big Bang point of our universe.

10.2 Spheroidal Universes

If a universe has a uniform distribution of galaxies and other mass-energy, as is commonly assumed,, then the universe has a spherical 3-surface in a larger dimensional space such as the Megaverse. We now describe the proof of this assertion.

We note a universe has a well-defined energy in a Megaverse. The energy consists of an internal part $U_{internal}$ and a surface energy part $U_{surface}$.

$$U = U_{internal} + U_{surface}$$

Using the principle of virtual displacements one can calculate surface energy contributions for infinitesimal surface elements δS. We see

$$\delta U_{surface} = -\delta W = \beta \delta S$$

where β is the surface tension, δS is an infinitesimal area element, and δW is the virtual work performed during the virtual displacement. Setting $\delta U_{surface} = 0$ and using Lagrange multipliers one arrives at the equation of a 3-surface sphere.

If the initial assumption of a uniform distribution of matter within the universe does not hold due to the 'nearess' of other universes, then the shape of the universe becomes spheroidal.

11. The Psychology of Physical Reality

11.1 Progress in Understanding of Physical Reality

In the past 2,500 years we have progressed in our understanding of the nature of matter and energy from the four elements theories of 500 B.C. to the atomic theory of matter (accepted[49] in only the late 19th Century) and the quantum photon theory of light.

In the 20th Century we saw the joint emergence of Relativity, Quantum Theory, and elementary particles and their interactions. Those developments led to the derivation of the Unified SuperStandard Model by the author at the beginning of the 21st Century (in the Book.)

The trend illustrated by these milestones ranges from the 'concrete' substances of 500 B.C. to the diaphanous wave-particles that we understand to constitute matter and energy at present. The theory that we have developed in these volumes for wave-particles lends itself to an interpretation as a language description. We have described the language description in chapter 8 and a series of earlier books. It has an alphabet of particles, and production rules embodied in the lagrangian of the Unified SuperStandard Model, that is particularly evident in the Model's perturbation theory expansion.

Thus one may say that, in 2,500 years, we have transitioned from 'concrete' matter to an 'evanescent' language of physical reality where matter and energy have the insubstantial form of a language.

11.2 Language and Psychology

Given this linguistic view of physical reality it is reasonable to inquire as to the 'psychology' that the Unified SuperStandard Model reflects. For a language reflects the culture and concerns of its speakers. The classic example of this 'hidden' aspect of languages is Ancient Greek, which had approximately three hundred words for the various forms of love suggesting a preoccupation for human emotions that we see in the dialogues of Plato and other Pre-Socratic Greek writings. Contrast that with the one word 'Love' that English uses for most forms of love.

In the case of the Unified SuperStandard Model we may expect to acquire physically relevant information about the physical psychology of the Entity but human-like aspects such as the emotions of love, hate, and so on cannot be determined from the consideration of the physical language of the Unified SuperStandard Model.

---

[49] One of the causes of Boltzmann's suicide was despair over the non-acceptance of atomic theory.

So what does the language of physical reality tell us of the underlying psychology of the putative Creator Entity?

1. It is abstract like mathematical reasoning..
2. It is computationally-oriented. The Unified SuperStandard Model can be mapped to a type 0 language and grammar production rules.
3. It is rigorously logical.
4. It is oriented toward simplicity in the manner of the Decision Axioms in the Book.

11.3 Entity Decision Axioms

For the convenience of the reader we reproduce below the Decision Axioms of Appendix C in the Book.

C.1.3 Decision Axioms for a Fundamental Theory

The concept of Decision Axioms does not seem to have been formally considered previously. Decision axioms guide a choice between alternatives in the construction of fundamental theory. Given the complex nature of fundamental theories choices between alternatives frequently occur. SuperString theory, for example, has millions of choices in the determination of the 'correct' SuperString theory.

What rules (Decision Axioms) can guide a choice? In this subsection we identify eight rules that provide guidance for making the proper choice.

C.1.3.1 Ockham's Razor

William of Ockham proposed a Law of Parsimony that is called *Ockham's Razor* which states that the simplest choice is to be preferred in a multiple choice situation. This principle is often stated as 'the simplest solution is usually the correct solution.'[50]

The best rationale for this principle is that it generally reduces the complexity that follows such a choice. Since many physics calculations are extremely difficult, picking the simplest choice.would generally tend to make subsequent theorems/and calculations less difficult. This point of view might be thought to be ad hoc or anthropomorphic. But it reflects the reality of scientific calculation and of theory construction.

Thus we will assume Ockham's Law of Parsimony in the construction of our theory with the proviso that Leibniz's Minimax Principle (below) takes precedence if there is a conflict in the implications of the choices.

C.1.3.2 Leibniz's Minimax Principle for Physics Theories

Leibniz[51] developed a Minimax Principle that can be phrased for our purposes as, "The universe is based on the smallest set of properties or features that lead to the greatest variety of phenomena." This principle reflects the spirit of the minimum/maximum criteria of the Calculus of Variations[52] that plays a central role in many physics theories. This principle somewhat overlaps Ockham's Law of Parsimony. Given a choice of possible theoretical lines of construction there is a possibility that the Law of Parsimony and the Minimax Principle would suggest different

---

[50] William of Ockham – Law of Parsimony – "Pluralitas non est ponenda sine necessitate" or "Plurality should not be posited without necessity." Ockham's Law was first stated by Durand De Saint-Pourçain (1270-1334 A.D.). In simple terms the principle states the simplest solution to a problem is most likely to be the correct solution.
[51] See Rescher (1967).
[52] Leibniz was one of the founders of the Calculus of Variations.

choices. Fortunately, we will see that the construction of the SuperStandard Model does not seem to present this potential dilemma.

An important, unremarked, aspect of Leibniz's Principle is the decision between a set of choices depends on the future part of the construction or theory. Thus future constructs influence past constructs in minimax decisions.[53]

C.1.3.3 Conputationally Minimal

Given a choice between two alternatives that appear equally likely from general considerations and the view of the preceding two Decision Axioms, the preferred choice is the one leading to the most minimal (simplest) computations in the full theory.

C.1.3.4 Experimentally Verifiable Directly or Indirectly

Given a choice between two alternatives, one of which is subject to experimental verification, directly or indirectly although perhaps in the future, and the other choice is not, then the first alternative is preferred.

C.1.3.5 Level of Rigor

Physics is mathematical in nature. Mathematics strives for rigor and will not be satisfied without rigorous results except for conjectures and speculations. Physics prefers rigorously derived results as wel. However, there is 'physical rigor,' which is a level of rigor that is rigorous to the extent that it is mathematically rigorous. This circuitous statement reflects the historical fact that Physics has often gone beyond the mathematics of the time. The most significant example is Newtonian physics, which until the mid-19th century used *derivatives* asserting that they were, or would become, rigorous mathematically. After a careful analysis of derivatives, and advances in the understanding of continuity and the various types of infinities, mathematicians were able to put derivatives on a rigorous footing justifying the use of derivatives by physicists for almost four centuries.

Today, the most interesting part of physics in search of mathematical rigor is the path integral formulation in quantum theory and the Faddeev-Popov Method, in particular. The problem here is again an understanding of infinities in functional derivatives and path integrals. Again, mathematical rigor is lackng. Yet physicists use these techniques with the strong belief that their results will be eventually justified rigorously.

In view of the nature of rigor in Physics it seems reasonable to state the Decision Axiom: *In the case of a situation where several possible approaches are possible, one should choose the approach that is most rigorous – should one exist. If all approaches are at an 'equal' level of rigor then the approach that appears 'most' physical should be chosen, taking account of the other Decision Axioms.*

C.1.3.6 Replication Principle - Nature Tends to Repeat Successful Strategies

In many branches of Science one sees that successful strategies and techniques are repeated in several areas – perhaps with some variations and changes. Consequently, when a physical phenomena is similar to another physical phenomena, and all other axioms that were stated above are not relevant for the phenomena, then one should model the new phenomena in a manner similar to the successfully modeled phenomena.

---

[53] The knowledgable reader will remember Feynman's speculation that the physical universe may be evolving from the future into the past. Quantum field theories support such an interpretation of their mathematics. The similarity of this fundamental minimax principle's feature with the corresponding feature of physics theories encourages support for the minimax principle as a 'design' law of fundamental physical theory.

C.1.3.7 Principle of Maximal Serendipity

A correct physical theory will evolve from fundamental axioms with maximal simplicity to attain better-than-expected ends. [Our Unified SuperStandard Model exemplifies serendipity in certain parts of its derivation.]

C.1.3.8 Principle of Maximal Descent to a Complete Theoryy

The development from axioms to a 'complete' theory will take place maximally in the sense that the steps of the derivation lead rapidly to the theory. This principle is analogous to the commonly used mathematical principle of maximal descent.

11.4 Principle of Vitamorphism

The various coupling constants and mass values embodied in the Unified SuperStandard Model suggest a profound Vitamorphism on the part of the Entity. For small changes in these values could make life, in all the forms that we can envision, impossible.

Thus we can say the Entity 'designed' the Cosmos for life.[54]

11.5 Wave-Particles vs. Thought

The wave-particles that are at the basis of the Unified SuperStandard Model are very much like thoughts within a brain. If wave-particles experienced no interactions, then they would be just as insubstantial as brain waves—the only significant difference being the confinement of thoughts to the brain. So we can make the simple diagram:

Reality = Thoughts + Interactions

if we regard an interactionless form of the Unified SuperStandard Model as thought.

---

[54] If a deeper theory can fully justify the values of these constants so that the Entity had no other choice, then the motive of Vitamorphism is removed.

12. Proposals for the Future of Elementary Particle Theory and Experiment

12.1 Theory

The primary theoretical issue facing the Unified SuperStandard Model is the extension of the theory to yield the coupling constants and the masses of its elementary fermions and vector bosons. currently there is no theoretical framework available to fully determine their values.

12.2 Experiment Proposals

There are a number of major questions confronting experiment in view of the success of the Unified SuperStandard Model's determination of the structure of the theory of elementary particles.

12.2.1 Speed of Light

A decisive neutrino experiment should be performed to accurately measure the speed of neutrinos. Existing experiments suggest that neutrinos move faster than the speed of light. However their results are not conclusive. Resolving this question is of great importance not only for the Unified SuperStandard Model but also for the future expansion of Mankind to the stars.

Einstein's statement on this matter—properly stated—is: particles cannot accelerate up to a speed greater than the speed of light with real-valued accelerations and speeds.

This statement does not exclude particles with complex-valued accelerations and speeds accelerating from below to speeds in excess of the speed of light. It also does not exclude particles such as neutrinos and down-type quarks from always having speeds in excess of the speed of light.

12.2.2 Extremely Massive Particles

The Unified SuperStandard Model predicts a fourth generation of fermions. It also predicts three additional layers of four generations each. None of these very massive particles have been found. In addition the theory predicts four layers of vector bosons. Only one layer is partially known at present. There is an important need for much more powerful accelerators to search for these fermions and bosons.

12.2.3 Dark Particles

Dark particles have not as yet been found. However indirect astrophysical data[55] suggests they exist in greater quantity than 'normal' particles. Again more powerful accelerators may find evidence of them. They are predicted in some detail in the Unified SuperStandard Model.

12.2.4 Left-Right Imaginary Momenta of Quarks

Chapter 2 of the Book specifies complex-valued momenta for up-type and down-type quarks. In ultra-high energy experiments at CERN and other laboratories quark-gluon plasmas are generated. Within the quark-gluon plasmas, quarks exist with a distribution of momenta. If one could 'filter' the quark-gluon plasma to create a stream of quarks with the imaginary part of their momenta pointing in the same direction, then one could imagine making a starship engine with complex acceleration and velocity that would enable a starship to travel faster than the speed of light vastly shortening the time of travel to the stars.[56]

Experiments on imaginary momentum filtering of quark-gluon plasmas could have important consequences for interstellar travel.

---

[55] Chapter 14 of the Book shows our theory predicts the relative abundances of Dark energy and Dark matter in relatively close agreement with astrophysical data.
[56] See Blaha (2014b) and (2014c).

REFERENCES

Akhiezer, N. I., Frink, A. H. (tr), 1962, *The Calculus of Variations* (Blaisdell Publishing, New York, 1962).

Bjorken, J. D., Drell, S. D., 1964, *Relativistic Quantum Mechanics* (McGraw-Hill, New York, 1965).

Bjorken, J. D., Drell, S. D., 1965, *Relativistic Quantum Fields* (McGraw-Hill, New York, 1965).

Blaha, S., 1998, *Cosmos and Consciousness 2ndEdition*   (Pingree-Hill Publishing, Auburn, NH, 2003).

_____, 2002, *A Finite Unified Quantum Field Theory of the Elementary Particle Standard Model and Quantum Gravity Based on New Quantum Dimensions™ & a New Paradigm in the Calculus of Variations* (Pingree-Hill Publishing, Auburn, NH, 2002).

_____, 2003, *A Finite Unified Quantum Field Theory of the Elementary Particle Standard Model and Quantum Gravity Based on New Quantum Dimensions™ and a New Paradigm in the Calculus of Variations* (Pingree-Hill Publishing, Auburn, NH, 2003).

_____, 2004, *Quantum Big Bang Cosmology: Complex Space-time General Relativity, Quantum Coordinates™Dodecahedral Universe, Inflation, and New Spin 0, ½, 1 & 2 Tachyons & Imagyons* (Pingree-Hill Publishing, Auburn, NH, 2004).

_____, 2005a, *Quantum Theory of the Third Kind: A New Type of Divergence-free Quantum Field Theory Supporting a Unified Standard Model of Elementary Particles and Quantum Gravity based on a New Method in the Calculus of Variations* (Pingree-Hill Publishing, Auburn, NH, 2005).

_____, 2005b, *The Metatheory of Physics Theories, and the Theory of Everything as a Quantum Computer Language* (Pingree-Hill Publishing, Auburn, NH, 2005).

_____, 2005c, *The Equivalence of Elementary Particle Theories and Computer Languages: Quantum Computers, Turing Machines, Standard Model, Superstring Theory, and a Proof that Gödel's Theorem Implies Nature Must Be Quantum* (Pingree-Hill Publishing, Auburn, NH, 2005).

_____, 2006a, *The Foundation of the Forces of Nature* (Pingree-Hill Publishing, Auburn, NH, 2006).

_____, 2006b, *A Derivation of ElectroWeak Theory based on an Extension of Special Relativity; Black Hole Tachyons; & Tachyons of Any Spin.* (Pingree-Hill Publishing, Auburn, NH, 2006).

_____, 2007a, *Physics Beyond the Light Barrier: The Source of Parity Violation, Tachyons, and A Derivation of Standard Model Features* (Pingree-Hill Publishing, Auburn, NH, 2007).

_____, 2007b, *The Origin of the Standard Model: The Genesis of Four Quark and Lepton Species, Parity Violation, the ElectroWeak Sector, Color SU(3), Three Visible Generations of Fermions, and One Generation of Dark Matter with Dark Energy* (Pingree-Hill Publishing, Auburn, NH, 2007).

_____, 2008a, *A Direct Derivation of the Form of the Standard Model From GL(16) (Pingree-Hill Publishing, Auburn, NH, 2008).*

_____, 2008b, *A Complete Derivation of the Form of the Standard Model With a New Method to Generate Particle Masses Second Edition* (Pingree-Hill Publishing, Auburn, NH, 2008)

_____, 2009, *The Algebra of Thought & Reality: The Mathematical Basis for Plato's Theory of Ideas, and Reality Extended to Include A Priori Observers and Space-Time Second Edition* (Pingree-Hill Publishing, Auburn, NH, 2009).

_____, 2010a, *Operator Metaphysics: A New Metaphysics Based on a New Operator Logic and a New Quantum Operator Logic that Lead to a Mathematical Basis for Plato's Theory of Ideas and Reality* (Pingree-Hill Publishing, Auburn, NH, 2010).

_____, 2010b, *The Standard Model's Form Derived from Operator Logic, Superluminal Transformations and GL(16)* (Pingree-Hill Publishing, Auburn, NH, 2010).

_____, 2010c, *SuperCivilizations: Civilizations as Superorganisms* (McMann-Fisher Publishing, Auburn, NH, 2010).

_____, 2011a, *21st Century Natural Philosophy Of Ultimate Physical Reality* (McMann-Fisher Publishing, Auburn, NH, 2011).

_____, 2011b, *All the Universe! Faster Than Light Tachyon Quark Starships & Particle Accelerators with the LHC as a Prototype Starship Drive Scientific Edition* (Pingree-Hill Publishing, Auburn, NH, 2011).

_____, 2011c, *From Asynchronous Logic to The Standard Model to Superflight to the Stars* (Blaha Research, Auburn, NH, 2011).

_____, 2012a, *From Asynchronous Logic to The Standard Model to Superflight to the Stars volume 2: Superluminal CP and CPT, U(4) Complex General Relativity and The Standard Model, Complex Vierbein General Relativity, Kinetic Theory, Thermodynamics* (Blaha Research, Auburn, NH, 2012).

_____, 2012b, *Standard Model Symmetries, And Four And Sixteen Dimension Complex Relativity; The Origin Of Higgs Mass Terms* (Blaha Reasearch, Auburn, NH, 2012).

_____, 2013a, *Multi-Stage Space Guns, Micro-Pulse Nuclear Rockets, and Faster-Than-Light Quark-Gluon Ion Drive Starships* (Blaha Research, Auburn, NH, 2013).

_____, 2013b, *The Bridge to Dark Matter; A New Sister Universe; Dark Energy; Inflatons; Quantum Big Bang; Superluminal Physics; An Extended Standard Model Based on Geometry* (Blaha Reasearch, Auburn, NH, 2013).

_____, 2014a, *Universes and Megaverses: From a New Standard Model to a Physical Megaverse; The Big Bang; Our Sister Universe's Wormhole; Origin of the Cosmological Constant, Spatial Asymmetry of the Universe, and its Web of Galaxies; A Baryonic Field between Universes and Particles; Megaverse Extended Wheeler-DeWitt Equation* (Blaha Reasearch, Auburn, NH, 2014).

_____, 2014b, *All the Megaverse! Starships Exploring the Endless Universes of the Cosmos Using the Baryonic Force* (Blaha Research, Auburn, NH, 2014).

_____, 2014c, *All the Megaverse! II Between Megaverse Universes: Quantum Entanglement Explained by the Megaverse Coherent Baryonic Radiation Devices – PHASERs Neutron Star Megaverse Slingshot Dynamics Spiritual and UFO Events, and the Megaverse Microscopic Entry into the Megaverse* (Blaha Research, Auburn, NH, 2014).

_____, 2015a, *PHYSICS IS LOGIC PAINTED ON THE VOID: Origin of Bare Masses and The Standard Model in Logic, U(4) Origin of the Generations, Normal and Dark Baryonic Forces, Dark Matter, Dark Energy, The Big Bang, Complex General Relativity, A Megaverse of Universe Particles* (Blaha Research, Auburn, NH, 2015).

_____, 2015b, *PHYSICS IS LOGIC Part II: The Theory of Everything, The Megaverse Theory of Everything, U(4)⊗U(4) Grand Unified Theory (GUT), Inertial Mass = Gravitational Mass, Unified Extended Standard Model and a New Complex General Relativity with Higgs Particles, Generation Group Higgs Particles* (Blaha Research, Auburn, NH, 2015).

_____, 2015c, *The Origin of Higgs ("God") Particles and the Higgs Mechanism: Physics is Logic III, Beyond Higgs – A Revamped Theory With a Local Arrow of Time, The Theory of Everything Enhanced, Why Inertial Frames are Special, Universes of the Mind* (Blaha Research, Auburn, NH, 2015).

_____, 2015d, *The Origin of the Eight Coupling Constants of The Theory of Everything: U(8) Grand Unified Theory of Everything   (GUTE), S^8 Coupling Constant Symmetry, Space-Time Dependent Coupling Constants, Big Bang Vacuum Coupling Constants, Physics is Logic IV* (Blaha Research, Auburn, NH, 2015).

_____, 2016a, *New Types of Dark Matter, Big Bang Equipartition, and A New U(4) Symmetry in the Theory of Everything: Equipartition Principle for Fermions, Matter is 83.33% Dark, Penetrating the Veil of the Big Bang, Explicit QFT Quark Confinement and Charmonium, Physics is Logic V* (Blaha Research, Auburn, NH, 2016).

_____, 2016b, *The Periodic Table of the 192 Quarks and Leptons in The Theory of Everything: The U(4) Layer Group, Physics is Logic VI* (Blaha Research, Auburn, NH, 2016).

_____, 2016c, *New Boson Quantum Field Theory, Dark Matter Dynamics, Dark Matter Fermion Layer Mixing, Genesis of Higgs Particles, New Layer Higgs Masses, Higgs Coupling Constants, Non-Abelian Higgs Gauge Fields, Physics is Logic VII* (Blaha Research, Auburn, NH, 2016).

_____, 2016d, *Unification of the Strong Interactions and Gravitation: Quark Confinement Linked to Modified Short-Distance Gravity; Physics is Logic VIII* (Blaha Research, Auburn, NH, 2016).

_____, 2016e, *MoND: Unification of the Strong Interactions and Gravitation II, Quark Confinement Linked to Large-Scale Gravity, Physics is Logic IX* (Blaha Research, Auburn, NH, 2016).

_____, 2016f, *CQ Mechanics: A Unification of Quantum & Classical Mechanics, Quantum/Semi-Classical Entanglement, Quantum/Classical Path Integrals, Quantum/Classical Chaos* (Blaha Research, Auburn, NH, 2016).

_____, 2016g, *GEMS: Unified Gravity, ElectroMagnetic and Strong Interactions: Manifest Quark Confinement, A Solution for the Proton Spin Puzzle, Modified Gravity on the Galactic Scale* (Pingree Hill Publishing, Auburn, NH, 2016).

_____, 2016h, *Unification of the Seven Boson Interactions based on the Riemann-Christoffel Curvature Tensor* (Pingree Hill Publishing, Auburn, NH, 2016).

_____, 2017a, *Unification of the Eleven Boson Interactions based on 'Rotations of Interactions'* (Pingree Hill Publishing, Auburn, NH, 2017).

_____, 2017b, *The Origin of Fermions and Bosons, and Their Unification* (Pingree Hill Publishing, Auburn, NH, 2017).

_____, 2017c, *Megaverse: The Universe of Universes* (Pingree Hill Publishing, Auburn, NH, 2017).

_____, 2017d, *SuperSymmetry and the Unified SuperStandard Model* (Pingree Hill Publishing, Auburn, NH, 2017).

_____, 2017e, *From Qubits to the Unified SuperStandard Model with Embedded SuperStrings: A Derivation* (Pingree Hill Publishing, Auburn, NH, 2017).

_____, 2017f, *The Unified SuperStandard Model in Our Universe and the Megaverse: Quarks, ... ,* (Pingree Hill Publishing, Auburn, NH, 2017).

Eddington, A. S., 1952, *The Mathematical Theory of Relativity* (Cambridge University Press, Cambridge, U.K., 1952).

Fant, Karl M., 2005, *Logically Determined Design: Clockless System Design With NULL Convention Logic* (John Wiley and Sons, Hoboken, NJ, 2005).

Feinberg, G. and Shapiro, R., 1980, *Life Beyond Earth: The Intelligent Earthlings Guide to Life in the Universe* (William Morrow and Company, New York, 1980).

Gelfand, I. M., Fomin, S. V., Silverman, R. A. (tr), 2000, *Calculus of Variations* (Dover Publications, Mineola, NY, 2000).

Giaquinta, M., Modica, G., Souchek, J., 1998, *Cartesian Coordinates in the Calculus of Variations* Volumes I and II (Springer-Verlag, New York, 1998).

Giaquinta, M., Hildebrandt, S., 1996, *Calculus of Variations* Volumes I and II (Springer-Verlag, New York, 1996).

Gradshteyn, I. S. and Ryzhik, I. M., 1965, *Table of Integrals, Series, and Products* (Academic Press, New York, 1965).

Heitler, W., 1954, *The Quantum Theory of Radiation* (Claendon Press, Oxford, UK, 1954).

Huang, Kerson, 1992, *Quarks, Leptons & Gauge Fields 2nd Edition* (World Scientific Publishing Company, Singapore, 1992).

Jost, J., Li-Jost, X., 1998, *Calculus of Variations* (Cambridge University Press, New York, 1998).

Kaku, Michio, 1993, *Quantum Field Theory*, (Oxford University Press, New York, 1993).

Kirk, G. S. and Raven, J. E., 1962, *The Presocratic Philosophers* (Cambridge University Press, New York, 1962).

Landau, L. D. and Lifshitz, E. M., 1987, *Fluid Mechanics 2ⁿᵈ Edition*, (Pergamon Press, Elmsford, NY, 1987).

Misner, C. W., Thorne, K. S., and Wheeler, J. A., 1973, *Gravitation* (W. H. Freeman, New York, 1973).

Rescher, N., 1967, *The Philosophy of Leibniz* (Prentice-Hall, Englewood Cliffs, NJ, 1967).

Rieffel, Eleanor and Polak, Wolfgang, 2014, *Quantum Computing* (MIT Press, Cambridge, MA, 2014).

Riesz, Frigyes and Sz.-Nagy, Béla, 1990, *Functional Analysis* (Dover Publications, New York, 1990).

Sagan, H., 1993, *Introduction to the Calculus of Variations* (Dover Publications, Mineola, NY, 1993).

Sakurai, J. J., 1964, *Invariance Principles and Elementary Particles* (Princeton University Press, Princeton, NJ, 1964).

Streater, R. F. and Wightman, A. S., 2000, *PCT, Spin, Statistics, and All That* (Princeton University Press, Princeton, NJ 2000).

Weinberg, S., 1972, *Gravitation and Cosmology* (John Wiley and Sons, New York, 1972).

Weinberg, S., 1995, *The Quantum Theory of Fields Volume I* (Cambridge University Press, New York, 1995).

Weinberg, S., 2000, *The Quantum Theory of Fields Volume III Supersymmetry* (Cambridge University Press, New York, 2000).

Weyl, H., 1950, *Space, Time, Matter* (Dover, New York, 1950).

Weyl, H., (Tr. S. Pollard et al), 1987, *The Continuum* (Dover Publications, New York, 1987).

INDEX

Anthropic Principle, 17, 18
Asynchronous Logic, iv, 59, 60, 67
axioms, 20, 22, 53, 54
baryonic force, 67
Big Bang,58, 60, 61, 66
Bjorken, J. D., 58
Black Hole, 59
Blaha, 58, 65
Calculus of Variations, 58, 62, 63
Charmonium, iv, 66
Chomsky, 10, 17, 20, 23, 44, 45
Complex General Relativity, iv, 60
Cosmic Pseudo-Euclid Entity, 16
Cosmological Constant, 60
Dark Energy, 59
Decision Axioms, 22, 53, 54
divergences, 65
ElectroWeak, 59
Entity, 16, 52, 53, 55
Euclid, 20
Functional space, 22
Functional-Lagrangians, 42, 43
Functionals, 30
Generation Group, 60
Grammar, 17, 21
Higgs Mechanism, 60, 66
interactions, 66
Landauer mass, 29, 30
Leibniz, 53, 63
Megaverse, iv

Minimax Principle, 53
Newtonian physics, 54
nonterminal, 17, 44, 45
Ockham's Law of Parsimony, 53
Ockham's Razor, 53
Parallel Processes, 21
Parity Violation, 59
physical rigor, 54
primitive terms, 20, 21
Production Rules, 17, 21, 43, 45, 46, 47
quantum computers, 65
Quantum Entanglement, 21, 31, 60
quark, 65
quba, 34, 42
qube, 22, 29, 30, 31, 34, 43
Robertson-Walker metric, 65
Skeleton Functional Lagrangians, 42
Special Relativity, 59
spin, 66
Spookiness, i
Standard Model, 58, 65
SU(3), 59
SuperStandard Model, 62
symbols, 17, 23, 44, 45
terminal, 17, 44, 45
Vitamorphic, 17
Vitamorphic Principle, 18
Wave Space, i, 40, 41
wave-particle duality, 10, 27, 28

About the Author

Stephen Blaha is a well known Physicist and Man of Letters with interests in Science, Society and civilization, the Arts, and Technology. He had an Alfred P. Sloan Foundation scholarship in college. He received his Ph.D. in Physics from Rockefeller University. He has served on the faculties of several major universities. He was also a Member of the Technical Staff at Bell Laboratories, a manager at the Boston Globe Newspaper, a Director at Wang Laboratories, and President of Blaha Software Inc and of Janus Associates Inc. (NH).

Among other achievements he was a co-discoverer of the "r potential" for heavy quark binding developing the first (and still the only demonstrable) non-abelian gauge theory with an "r" potential; first suggested the existence of topological structures in superfluid He-3; first proposed Yang-Mills theories would appear in condensed matter phenomena with non-scalar order parameters; first developed a grammar-based formalism for quantum computers and applied it to elementary particle theories; first developed a new form of quantum field theory without divergences (thus solving a major 60 year old problem that enabled a unified theory of the Standard Model and Quantum Gravity without divergences to be developed); first developed a formulation of complex General Relativity based on analytic continuation from real space-time; first developed a generalized non-homogeneous Robertson-Walker metric that enabled a quantum theory of the Big Bang to be developed without singularities at t = 0; first generalized Cauchy's theorem and Gauss' theorem to complex, curved multi-dimensional spaces; received Honorable Mention in the Gravity Research Foundation Essay Competition in 1978; first developed a physically acceptable theory of faster-than-light particles; first derived a composition of extrema method in the Calculus of Variations; first quantitatively suggested that inflationary periods in the history of the universe were not needed; first proved Gödel's Theorem implies Nature must be quantum; provided a new alternative to the Higgs Mechanism, and Higgs particles, to generate masses; first showed how to resolve logical paradoxes including Gödel's Undecidability Theorem by developing Operator Logic and Quantum Operator Logic; first developed a quantitative harmonic oscillator-like model of the life cycle, and interactions, of civilizations; first showed how equations describing superorganisms also apply to civilizations. A recent book shows his theory applies successfully to the past 14 years of history and to new archaeological data on Andean and Mayan civilizations as well as Early Anatolian and Egyptian civilizations.

He first developed an axiomatic derivation of the forms of The Standard Model from geometry – space-time properties – The Extended Standard Model. It has a Dark Matter sector that approximates the ElectroWeak sector with Dark doublets and Dark gauge interactions. It also uses quantum coordinates to remove infinities that crop up in most interacting quantum field theories and additionally to remove the infinities that appear in the Big Bang and generate inflationary growth of the universe. The Extended Standard Model is expanded into The Extended SuperStandard Model presented in this volume andvolume 1.

Blaha has had a major impact on a succession of elementary particle theories: his Ph.D. thesis (1970), and papers, showed that quantum field theory calculations to all orders in ladder approximations could not give scaling deep inelastic electron-nucleon scattering. He later showed the eigenvalue equation for the fine structure constant α in Johnson-Baker-Willey QED had a zero at $\alpha = 1$ not $1/137$ by solving the Schwinger-Dyson equations to all orders in an approximation that agreed with exact results to 4th order in α thus ending interest in this theory. In 1979 at Prof. Ken Johnson's (MIT) suggestion he calculated the proton-neutron mass difference in the MIT bag model and found the result had the wrong sign reducing interest in the bag model. These results all appear in Physical Review papers. In the 2000's he repeatedly pointed out the shortcomings of SuperString theory and showed that The Standard Model's form could be derived from space-time geometry by an extension of Lorentz transformations to faster than light transformations. This deeper space-time basis greatly increases the possibility that it is part of THE fundamental theory.Recently, Blaha showed that the Weak interactions differed significantly from the Strong, electromagnetic and gravitation interactions in important respects while these interactions had similar features, and suggested that ElectroWeak theory, which is essentially a glued union of the Weak interactions and Electromagnetism, possibly modulo unknown Higgs particle features, be replaced by a unified theory of the other interactions combined with a stand-alone Weak interaction theory. Blaha also showed that, if Charmonium calculations are taken seriously, the Strong interaction coupling constant is only a factor of five larger than the electromagnetic coupling constant, and thus Strong interaction perturbation theory would make sense and yield physically meaningful results.

In graduate school (1965-71) he wrote substantial papers in elementary particles and group theory: The Inelastic E- P Structure Functions in a Gluon Model. Phys. Lett. B40:501-502,1972; Deep-Inelastic E-P Structure Functions In A Ladder Model With Spin 1/2 Nucleons, Phys.Rev. D3:510-523,1971; Continuum Contributions To The Pion Radius, Phys. Rev. 178:2167-2169,1969; Character Analysis of U(N) and SU(N), J. Math. Phys. 10, 2156 (1969); and The Calculation of the Irreducible Characters of the Symmetric Group in Terms of the Compound Characters, (Published as Blaha's Lemma in D. E. Knuth's book: *The Art of Computer Programming Vols. 1 – 4*).

In the early 1980's Blaha was also a pioneer in the development of UNIX for financial, scientific and Internet applications: benchmarked UNIX versions showing that block size was critical for UNIX performance, developing financial modeling software, starting database benchmarking comparison studies, developing Internet-like UNIX networking (1982) and developing a hybrid shell programming technique (1982) that was a precursor to the PERL programming language. He was also the manager of the AT&T ten-year future products development database. His work helped lead to commercial UNIX on computers such as Sun Micros, IBM AIX minis, and Apple computers.

In the 1980's he pioneered the development of PC Desktop Publishing on laser printers. and was nominated for three "Awards for Technical Excellence" in 1987 by PC Magazine for PC software products that he designed and developed.

Recently he has developed a theory of Megaverses – actual universes of which our universe is one – with quantum particle-like properties based on the Wheeler-DeWitt equation of Quantum Gravity. He has developed a theory of a baryonic force, which had been conjectured many years ago, and estimated the strength of the force based on discrepancies in measurements of the gravitational constant G. This force, operative in D-dimensinal space, can be used to escape from our universe in "uniships" which are the equivalent of the faster-than-light starships proposed in the author's earlier books. Thus travel to other universes, as well as to other stars is possible.

Blaha also considered the complexified Wheeler-DeWitt equation and showed that its limitation to real-valued coordinates and metrics generated a Cosmological Constant in the Einstein equations.

The author has also recently written a series of books on the serious problems of the United States and their solution as well as a book on the decline of Mankind that will follow from current social and genetic trends in Mankind.

In the past twelve years Dr. Blaha has written over 40 books on a wide range of topics. Some recent major works are: *From Asynchronous Logic to The Standard Model to Superflight to the Stars, All the Universe!, SuperCivilizations: Civilizations as Superorganisms, America's Future: an Islamic Surge, ISIS, al Qaeda, World Epidemics, Ukraine, Russia-China Pact, US Leadership Crisis,The Rises and Falls of Man – Destiny – 3000 AD: New Support for a Superorganism MACRO-THEORY of CIVILIZATIONS From CURRENT WORLD TRENDS and NEW Peruvian, Pre-Mayan, Mayan, Anatolian, and Early Egyptian Data, with a Projection to 3000 AD,* and *Mankind in Decline: Genetic Disasters, Human-Animal Hybrids, Overpopulation, Pollution, Global Warming, Food and Water Shortages, Desertification, Poverty, Rising Violence, Genocide, Epidemics, Wars, Leadership Failure.*

He has taught approximately 4,000 students in undergraduate, graduate, and postgraduate corporate education courses primarily in major universities, and large companies and government agencies.

The above paragraphs summarize much of his work over the past fifty years. This work is fully documented. He continues to engage in research and writing at Blaha Research.

www.ingramcontent.com/pod-product-compliance
Lightning Source LLC
Chambersburg PA
CBHW082009190326
41458CB00010B/3131